ENDING
PLASTIC WASTE

COMMUNITY ACTIONS AROUND THE WORLD

EDITORS

**BRITTA DENISE HARDESTY, KATHRYN WILLIS,
JUSTINE BARRETT** AND **CHRIS WILCOX**

CSIRO
PUBLISHING

A catalogue record for this book is available from the National Library of Australia.

ISBN: 9781486312290 (pbk)
ISBN: 9781486312306 (epdf)
ISBN: 9781486312313 (epub)

Published in print in Australia and New Zealand, and in all other formats throughout the world, by CSIRO Publishing.

CSIRO Publishing
Private Bag 10
Clayton South VIC 3169
Australia

Telephone: +61 3 9545 8400
Email: publishing.sales@csiro.au
Website: www.publish.csiro.au
Sign up to our email alerts: publish.csiro.au/earlyalert

A catalogue record for this book is available from the British Library, London, UK.

Published in print only, throughout the world (except in Australia and New Zealand), by CABI.

ISBN 9781800623613

CABI	CABI
Nosworthy Way	We Work
Wallingford	One Lincoln Street, 24th Floor
Oxfordshire OX10 8DE	Boston, MA 02111
UK	USA
Tel: +44 (0)1491 832111	Tel: +1 (617)682-9015
Email: info@cabi.org	E-mail: cabi-nao@cabi.org
Website: www.cabi.org	

Front cover: photo by xalien/Shutterstock.com; map by JoelMasson/Shutterstock.com; recycle icon by prabath23/Shutterstock.com

Title page: map by JoelMasson/Shutterstock.com; recycle icon by prabath23/Shutterstock.com

Edited by Natalie Korszniak
Cover design by Cath Pirret
Typeset by Envisage Information Technology
Printed in Malaysia by Papercraft

CSIRO Publishing publishes and distributes scientific, technical and health science books, magazines and journals from Australia to a worldwide audience and conducts these activities autonomously from the research activities of the Commonwealth Scientific and Industrial Research Organisation (CSIRO). The views expressed in this publication are those of the author(s) and do not necessarily represent those of, and should not be attributed to, the publisher or CSIRO. The copyright owner shall not be liable for technical or other errors or omissions contained herein. The reader/user accepts all risks and responsibility for losses, damages, costs and other consequences resulting directly or indirectly from using this information.

CSIRO acknowledges the Traditional Owners of the lands that we live and work on and pays its respect to Elders past and present. CSIRO recognises that Aboriginal and Torres Strait Islander peoples in Australia and other Indigenous peoples around the world have made and will continue to make extraordinary contributions to all aspects of life including culture, economy and science. CSIRO is committed to reconciliation and demonstrating respect for Indigenous knowledge and science. The use of Western science in this publication should not be interpreted as diminishing the knowledge of plants, animals and environment from Indigenous ecological knowledge systems.

The paper this book is printed on is in accordance with the standards of the Forest Stewardship Council® and other controlled material. The FSC® promotes environmentally responsible, socially beneficial and economically viable management of the world's forests.

FOREWORD

The United Nations global approach and regional action to reducing plastic pollution

After decades of increasing plastic production, consumption and mismanaging plastic waste, we are now aware of the consequences that our plastic reliance has on our oceans, ecosystems and food chains. Our historically linear economy for plastic means that much of the plastic produced is used only once before ending up in landfill or escaping into our natural environments. The use of personal protective equipment and additional packaging brought on by the COVID-19 pandemic has placed additional pressure on the global plastic pollution problem. Because plastic pollution does not stop at borders and value chains are globalised and complex, we need combined efforts across the globe and across stakeholders to end plastic pollution. Only through collaboration will we achieve success.

At a global level, the United Nations Environment Programme (UNEP) has spearheaded several strategies to tackle transboundary plastic pollution and marine litter in the environment. Most recently, countries around the globe came together in the United Nations Environment Assembly in its Fifth Session to endorse a historic resolution to end plastic pollution by negotiating an international legally binding instrument by 2024. This instrument will consider interventions along the life cycle of plastic, including its production, design and disposal, and will address the technical assistance and financing needs, as well as access to science and data, for effective decision making. Member states have realised the urgency for collective action to change our relationship with plastic.

To accelerate action, we can build on existing regional tools and put regulatory frameworks and financing mechanisms in place that will bring about system changes. Regional Seas Conventions and Action Plans are examples of intergovernmental mechanisms that can be used for the coordinated protection of the marine environment. The UNEP Ecosystems Division administers Regional Seas programs in the Caribbean, East Asian Seas, Eastern Africa, Mediterranean, North-West Pacific and Western Africa regions. The UN provides governments and stakeholders in the region access to marine litter good practices, research, data and capacity building to promote evidence-based decision making.

Tackling marine litter has become a very crowded space. Governments can leverage existing regional mechanisms to overcome financial, technical, capacity-building and knowledge barriers identified at the global level and to coordinate coherent regional responses to transboundary challenges of marine litter.

One such example is the Coordinating Body on the Seas of East Asia (COBSEA), with its nine participating countries in South-East and East Asia. The COBSEA Regional Action Plan on Marine Litter (RAP MALI) recognises that monitoring and assessment are indispensable in identifying the status and trends in marine litter status, as well as its most critical effects. In partnership with the Commonwealth Scientific and Industrial Research Organisation (CSIRO), COBSEA developed Regional Guidance to harmonise and strengthen marine litter monitoring programs in the region for greater data comparability in line with global guidelines and national capacities. Monitoring marine litter plays a crucial role in tracking progress at various levels for effective interventions.

To help create change at a policy level, the UNEP has developed a Marine Litter Legislation toolkit for policy makers, and provides technical assistance and capacity building to replicate good practices and develop effective policies and regulations at a national level. The UNEP also recognises the role that bottom-up initiatives play in changing behaviour, creating awareness at local levels and influencing policy. Grassroots initiatives are often inspired, developed and run by local people to address a local pollution problem and improve local livelihoods. They can have a large impact on reducing pollution at relatively low costs, and can play an important role in the collection of on-the-ground data regarding the quantity and type of litter in the environment. This information is vital in advocating for change at governance, manufacturing and end-user levels.

Most of the organisations described in this book take a bottom-up approach and are a source of inspiration to others. Armed with knowledge and the stories told in this book, readers will feel inspired and empowered to create their own organisation and join us in our collective effort towards solving the global plastics pollution challenges.

Natalie Harms
Programme Officer on Marine Litter, Secretariat of the Coordinating Body on the Seas of East Asia (COBSEA), United Nations Environment Programme (UNEP)

CONTENTS

ACKNOWLEDGEMENTS

We thank the authors who contributed formal chapters to paint the picture of global pollution and waste management. We thank those who are involved in the programs described in this book that are contributing to improvements in waste management solutions around the world. We also thank Vanessa Mann and Leigh Keen for assisting in the early development and administration of this book. Finally, we thank the CSIRO Publishing team who patiently guided us through this process.

LIST OF CONTRIBUTORS

JAMES BAKER

James Baker is a senior circular economy specialist in the sustainable development and climate change department at the Asian Development Bank. He manages the regional marine plastics reduction program, supporting the Asian Development Bank (ADB) Healthy Oceans Action Plan and Operation Pillar 3. James is leading the ADB framework on a circular economy.

TRISH HYDE AND MURRAY HYDE

Having built and successfully piloted PlastX (their demand-driven, post-consumer recycled plastic-sourcing platform that engages waste collectors to recover material to optimise its circular economic value for brands), Trish and Murray Hyde are in the middle of preparations for scaling their organisation.

WINNIE LAU

With a strong background in guiding climate science and technology policy for government agencies, Winnie Lau is a project director working on the prevention of ocean plastic at The Pew Charitable Trusts. She also played a lead role in developing new projects and partnerships in Asia with Pew's International Conservation Unit.

COSTAS VELIS

Costas Velis is an international expert on the circular economy, resources from waste and plastics pollution. With over 18 years of practical experience in resources recovery, he serves on many international committees and initiatives addressing relevant challenges of global scale.

AZRA YAQUB VAWDA

Azra Yaqub Vawda is an independent development advisor who is involved in the development of strategy and in the roadmap for the Blue Finance Hub in the ASEAN region, which promotes action on plastic pollution in Asia and the Pacific regions. Her areas of expertise are financing the green and blue economies through innovative approaches.

ABOUT THIS BOOK

'Reducing waste around the world'

Welcome to this informal waste management guidebook: a collection of programs around the world that are turning waste into a resource, increasing the social capacity of their communities and reducing the amount of waste in their environment.

This book contains a collection of stories from diverse programs across many different countries that have been developed by people who are committed to helping manage waste within their communities. Each program is designed outside of the established formal waste management systems present in many Organisation for Economic Co-operation and Development (OECD) countries.

We present approaches of the various programs as a 'how to' guide for environmental ministers, public works officers and community groups who want to improve waste management and reduce litter in their community. Each program states how many people it can benefit and the resources required.

Each story is told in the words of the founders and creators of each program, and we thank them for the time they dedicated to answering our questions. We have also collated advice and information from experts in the fields of waste management, environmental finance and plastic pollution research to give a broader outlook on why the programs we have included in this book are so vital to protecting our planet.

Our aim with this book is to share the success stories of how informal organisations around the world are making a positive impact on both the environment and communities through their programs.

Above all, these stories are a tribute to the organisations, people and programs that have established their own waste management systems. We hope this book not only inspires others to start their own programs, but also provides guidance on what types of program could be used, how to successfully implement the program, the resources needed and the lessons learned to overcome barriers.

We dedicate this book to our partners and their programs. We hope you enjoy learning about each of them as much as we did.

Britta Denise Hardesty, Kathryn Willis, Justine Barrett and Chris Wilcox
Co-editors

1

Introduction

Britta Denise Hardesty, CSIRO, Australia, and
Justine Barrett, CSIRO, Australia

The genesis of this book comes from not only the recognition of plastic pollution as a growing social, economic, environmental and potential health issue, but also from participating in several international events over the past decade or more in which numerous solutions were put forth by governments, industry and important stakeholders from around the world. We have all seen over the past several years an increased awareness of the causes, consequences and patterns of destruction wrought by plastic pollution in the environment. At various international meetings there are discussions about infrastructure needs, resource requirements in the tens of millions and billions of dollars, political challenges and opportunities. There are also important discussions about the need for broad systemic change, and an acknowledgment of the complexity of the plastics pollution issue.

However, what is often missing from these conversations is a focus on and an acknowledgement of the incredible actions that are already being taken and are already making a difference, improving people's lives in the communities in which these new businesses have been created and where people have had the insight, initiative and commitment to make something happen, despite limited resources and other potential obstacles standing in their way.

These are exciting times, with the United Nations Environment Assembly launching negotiations for an international legally binding instrument to end plastic pollution and establish a treaty by 2024. This treaty is anticipated to address the full life cycle of plastics, including plastic production, design and disposal. The ambitious time line given for this important resolution highlights the global appetite and commitment by members states to change our relationship with plastic at local, national, regional and global levels.

Many of us are familiar with the waste hierarchy (see figure on page 2). We know that we, as a global community, need to prevent or minimise the use of plastics in our daily lives, and we see ingenuity in communities around the world adopting novel approaches to using a variety of products that are purchased and either packaged in or made from plastic. These reuse and repurpose approaches are many and varied, and some of the products featured in this book are incredible examples of turning something that was previously considered waste into a commodity; into something that is newly created,

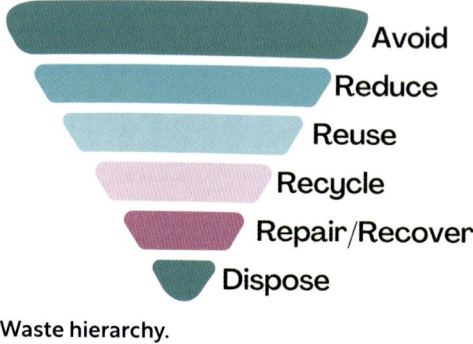

Waste hierarchy.

fit for purpose, with both value and an extended lifetime. This book also includes stories of ways to reduce landfill and waste disposal by recovering and repurposing materials to give them a new lease on life.

When we were identifying stories for inclusion in this book, given the current knowledge gaps in the important work being undertaken, we looked to ensure a variety of different key criteria. First, we focused on projects or programs from the informal and/or non-government sectors. We also focused on programs or activities that involved reusing, repurposing or recycling 'waste' products. We did not limit the programs to those only focusing on items that are commonly found in marine debris or coastal litter surveys, nor did we limit the programs to those that solely or explicitly focus on *marine* plastics, because we are conscious that most of what is considered 'marine debris' is typically land-based waste or coastal litter.

However, we did focus on including projects and activities that contribute to the economic and social welfare of disadvantaged communities. Acknowledging that ocean pollution and waste mismanagement are social equity issues is important, and so highlighting the exciting programs that have been developed that aim to improve the community, economic, health and social benefits of individuals and communities is important. Finally, we were particularly interested in identifying programs that look for integrated approaches, are holistic and come up with a local solution for a particular context but can potentially be scaled up and implemented or adapted to different contexts, communities and cultures to achieve benefits in other parts of the world.

We know that these programs are not the only stories of triumph taking place. However, we wanted to highlight those programs that can be adapted to suit the needs and waste issues of different countries and to share diverse stories from countries around the world of organisations that are addressing different types of plastic materials with various approaches and across an array of contexts.

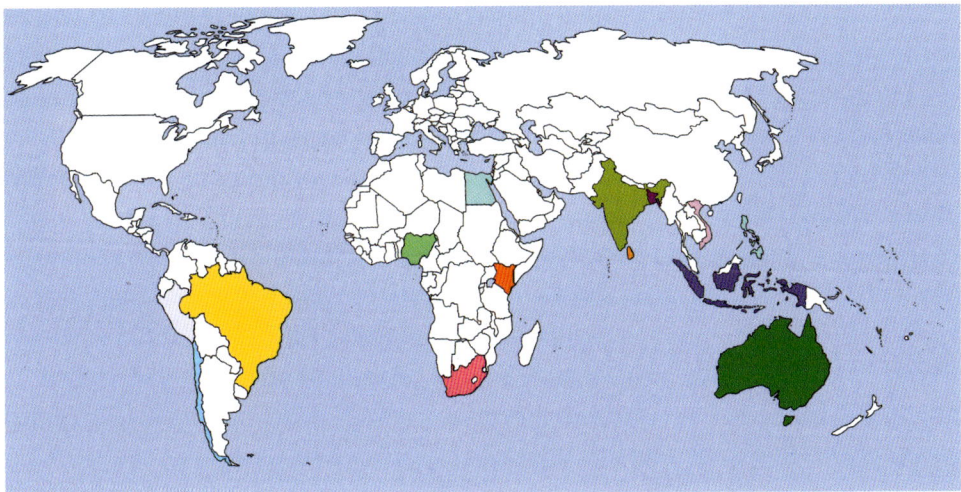

A world map highlighting the countries in which the organisations telling their stories in this book operate.

We are excited to share with you these stories of ingenuity, commitment and success that show how, where and why people around the world are fundamentally moving the dial: they are creating new products, identifying new markets and engaging with millions of people to help them change their plastic footprint and reconsider their relationship with plastic. Please read about these programs, these stories, with an eye for how they can be adapted in your community and where they may be relevant in your country, and perhaps use them to spark inspiration and action not only for yourself, but also for people, governments, industries and the world at large.

Plastic waste inputs and leakage: scale, drivers and solutions

Dr Winnie Lau, The Pew Charitable Trusts, USA

In this chapter, Winnie outlines the plastic pollution challenges our planet is facing. She paints a grim picture of what will happen if we do not make a collective effort to change our actions around plastic production, use and disposal. This chapter also highlights the value of the organisations and programs that work towards reducing plastic in our environment, many of which are proudly showcased in this book.

Plastic is ubiquitous. It has been found in all corners of the world: at the peaks of the highest mountains, in the depths of the Mariana Trench and in uninhabited Antarctica. Marine species are encountering plastics in these habitats, eating plastic, getting entangled in plastic or being impacted in some other way. People don't escape either: microplastics have been found in seafood and bottled water, in the human placenta of unborn babies[1] and in our blood streams.[2] The very properties that make plastic so useful also mean that once plastic enters the environment, it will remain there for hundreds of years or longer.

Plastic was found in the ocean as early as the 1970s. In 2015, Jambeck *et al.* were the first to estimate the ocean plastic problem at the global scale: 8 million metric tons of plastic (not including microplastics) entered the ocean from land in 2010, the equivalent of dumping one garbage truckload of plastic into the ocean every minute.[3]

Several proposed solutions followed, ranging from banning all plastic, to landfilling, incinerating or recycling plastic, to using alternative or new materials. However, there was little evidence as to which, if any, of these strategies could effectively stop plastic pollution globally, and within a feasible time frame.

To evaluate these various strategies and identify pathways towards stopping plastic pollution, The Pew Charitable Trusts and SYSTEMIQ, along with four thought partners and 17 global experts, conducted a global analysis and published the findings in the 2020 report *Breaking the Plastic Wave*.[4] The report confirmed the urgency to act.

In 2016, 11 million metric tons of plastic (including both macro- and microplastics) entered our oceans. If the world continues with 'business as usual', this would nearly triple to 29 million metric tons by 2040, an amount equivalent to dumping 50 kg of plastic waste along every metre of coastline globally.[4] As such, the amount of plastic accumulated in the ocean by 2040 would be approximately four times what is in the ocean today. The problem of plastic pollution on land is even more stark, rising at nearly double the rate of ocean pollution, with an estimated increase from 18 million metric tons per year in 2016 to 52 million metric tons in 2040.[5]

The *Breaking the Plastic Wave* report also evaluated three previously proposed solutions that focus on a specific part of the plastics system: collect and dispose; recycle; and reduce and substitute.[4] Even with ambitious action, each of these solutions would, at best, only achieve keeping plastic waste entering the oceans at around the 2016 level of 11 million metric tons per year. And this would still require considerable effort and investment.

The only scenario where a substantial decrease in plastic waste going into the ocean could be achieved would be to implement solutions across the entire plastic system, synergistically and ambitiously: a system change scenario. A system change approach includes first examining whether plastic is needed or is the best material for a particular purpose, followed by eliminating the plastic that is not needed (e.g. product redesign or reducing overpackaging). For plastic that is needed, designing materials that can be reused and recycled, and then recapturing as much of this material through recycling as possible, are the ultimate goals. Then, safely disposing of the remaining plastic to keep it from escaping into the environment is critical. This system change shows there is a pathway to cut plastic pollution going into the ocean by 80% by 2040.

There are four main drivers that are key to reducing plastic leakage into the environment: (1) shifting the dial on the projected rapid growth in plastic production and waste generation; (2) alleviating the growing collection gap in developing countries; (3) mitigating the increasing production and manufacturing of high-leakage plastics; and (4) revising the fundamental economics of the plastics system that are currently in place. Each of these drivers is discussed in more detail below.

A lack of formal waste collection services can lead to waste piling in communities.

Global plastic production is projected to double over the next two decades. It is estimated that in low- and middle-income countries waste generation per person will double from 20 to 38 kg.[4] Despite currently generating almost four times as much waste per person each year (76 kg), absolute waste generation per person per year in high-income countries is projected to grow similarly to an estimated 95 kg.[4]

Due to the rapid growth in plastic waste generation, there will be a growing collection gap in developing countries. It is estimated that 2 billion people today, concentrated mainly in low- and middle-income countries, do not have access to formal waste collection services (although this is improving, as you will read in some of the programs included in this book). Without significant investment in waste management infrastructure, the number of people without formal waste collection is expected to grow to 4 billion in 2040.[4]

Compounding the waste collection gap is the anticipated shift to low-value, hard-to-collect, hard-to-recycle plastic. Soft plastics, like plastic bags and plastic packaging/shrink wraps, and products that contain multiple types of plastic disproportionately leak to the environment. Combined, these two categories of plastics made up 60% of the production in 2016, but accounted for 80% of the plastic waste lost into the ocean.[4]

Underlying the above drivers is an imbalance in the economics of the current global plastics economy. Today, virgin plastic, in most cases, is cheaper than recycled plastic material. For items such as plastic bottles, where only one type of plastic is used (polyethylene terephthalate (PET)), recycling is economically viable, meaning that the cost of recycled bottles is on a par with virgin plastic. However, soft plastics and products containing multiple layered plastics are nearly impossible to recycle because they do not yield enough value to be collected and sorted for recycling. Not surprisingly, the global recycling rate for all plastics is estimated at only 15%.[4] Moreover, investments in plastic production outstrip those in waste management infrastructure by more than one order of magnitude.[4]

A key group that will play a significant role in overcoming the challenges associated with the drivers described above will be the informal waste collection sector in low- and middle-income countries. As you will read about in the programs throughout this book, as well as in the next chapter, globally, the informal waste collection sector, also called 'waste pickers', has outpaced formal waste management services in contributing to the global plastics recycling industry. However, waste

Beach cleaning operations. PHOTOGRAPH: SYLLA CHEICK 225/ATTRIBUTION-SHARE ALIKE 4.0 INTERNATIONAL (CC BY-SA 4.0): HTTPS://CREATIVECOMMONS.ORG/LICENSES/BY-SA/4.0/DEED.EN.

pickers have largely been unrecognised by governments or society, have been underpaid and often are working in unsanitary or unsafe conditions. Formally incorporating the informal waste collection sector into the plastic system could increase the overall global capacity to collect, sort and recycle plastic waste and, at the same time, contribute to equity and social justice.

From the economic and climate perspectives, there are also tangible incentives to prevent plastic waste generation and pollution. Due to the proliferation of single-use items and packaging that is thrown away after a short use, the global economy loses US\$80–120 billion dollars a year in plastic material.[6] Businesses globally may also face a financial risk of US\$100 billion dollars over the next 20 years if governments pass on the additional cost of waste management to them.[4] In addition, the greenhouse gas emissions associated with the plastic value chain are projected to increase by 2.5-fold to 2.1 gigatons of CO_2 equivalent per year by 2040 under the business-as-usual trajectory, and risk using up a significant share of the carbon budget allowable under the Paris Agreement.[4]

Plastic pollution is a complex issue. Solutions will require actions across all of society and through a systemic approach. These solutions and policies need to be put in place urgently and ambitiously. Significant investments will be needed in infrastructure, as well as in innovation, and the marginalised informal waste collection sector will need to be recognised and incorporated into the waste management sector. With bold action, the plastic pollution problem can be addressed within one generation.

References

1   Ragusa A, Svelato A, Santacroce C, Catalano P, Notarstefano V, *et al.* (2021) Plasticenta: first evidence of microplastics in human placenta. *Environment International* **146**, 106274. doi:10.1016/j.envint.2020.106274

2   Leslie HA, van Velzen MJM, Brandsma SH, Vethaak AD, Garcia-Vallejo JJ, *et al.* (2022) Discovery and quantification of plastic particle pollution in human blood. *Environment International* **163**, 107199. doi:10.1016/j.envint.2022.107199

3   Jambeck JR, Geyer R, Wilcox C, Siegler TR, Perryman M, *et al.* (2015) Plastic waste inputs from land into the ocean. *Science* **347**(6223), 768–771. doi:10.1126/science.1260352

4   The Pew Charitable Trusts, SYSTEMIQ (2020) *Breaking the Plastic Wave: A Comprehensive Assessment of Pathways Towards Stopping Ocean Plastic Pollution.* The Pew Charitable Trusts, USA, <https://www.pewtrusts.org/-/media/assets/2020/07/breakingtheplasticwave_report.pdf>.

5   Lau WWY, Shiran Y, Bailey RM, Cook E, Stuchtey MR, *et al.* (2020) Evaluating scenarios toward zero plastic pollution. *Science* **369**(6510), 1455–1461. doi:10.1126/science.aba9475

6   World Economic Forum, Ellen MacArthur Foundation, McKinsey & Co. (2016) *The New Plastics Economy: Rethinking the Future of Plastics.* Ellen MacArthur Foundation, UK, <https://ellenmacarthurfoundation.org/the-new-plastics-economy-rethinking-the-future-of-plastics>.

Waste pickers: keeping after-use plastics circular, but at high personal cost

Costas Velis, University of Leeds, UK

As you read about the wonderful programs in our book you will learn about the vital role played by the informal waste sector, predominantly made up of waste pickers, in recycling waste materials. In fact, this sector is responsible for over half the plastics recycled worldwide, not to mention other valuable materials collected and returned into the supply chain. In this chapter, Costas describes the valuable contributions of waste pickers all over the world, and the dangerous working conditions they face on a daily basis.

Who are the people making up the informal collection and recycling sector?

Catadores in Brazil, *Cartoneros* in Argentina and *Recicladores* in other Spanish-speaking countries, 'waste pickers' (as they call themselves collectively in English) are the tens of millions of people around the world working in the informal recycling sector (IRS) who make their livelihood by identifying, collecting, sorting and selling the most sought-after recyclable items from municipal solid waste. These items include paper, glass, aluminium and plastics.

These workers are 'informal' in the sense that, typically, their activity is not officially registered and they may not pay taxes or contribute to national insurance. These workers are part of the informal sector. This is not unusual across the Global South. These workers are also 'informal' in contrast with the 'formal' waste management industry. Beware: such informality has nothing to do with the level of these workers' skills or expertise, which they develop on the job. They must be highly

A waste picker collecting recycled materials at a temporary dumpsite, where waste is accumulated before collection, Nairobi. PHOTOGRAPH: COSTAS VELIS.

skilled to survive and, in fact, are experts in manually identifying the materials that our consumer products, including plastic packaging, are made of.

Waste pickers operate in a variety of places and roles. Some waste pickers operate across chaotic and dangerous dumpsites, risking their health, and sometimes their life, to reclaim waste that has been sent there at 'end of life'. Waste in these dumpsites is concentrated and accessible, making it a 'good' place to sort through and salvage materials. Other waste pickers operate on the streets, searching bin by bin, and others still service businesses and households as itinerant buyers of recyclables – there are endless variations. Waste pickers also range in age from children, who are forced to work prematurely, to pensioners, and you will find waste pickers in countries all around the world. That said, all waste pickers share a common core feature: they are socially vulnerable, in one or more ways. They are also physically vulnerable, having to manually handle waste items created in our modern society.

Environmentally most relevant – yet vulnerable

As expert collectors and recyclers, waste pickers provide much needed environmental services and, arguably, their activities have a strongly positive overall sustainability footprint. Waste pickers salvage waste items that, by and large, would otherwise be dumped or disposed of, possibly in environmentally damaging ways, in dumpsites or via open uncontrolled burning.

Waste pickers support a circular economy by enabling waste to be turned back into useful products. Waste pickers often operate where there is no recycling system set up by a state or formal industry; therefore, they contribute to the positive energy and materials savings achieved through recycling. Waste pickers supply the industry of the Global South with affordable recycled materials, in the meanwhile making themselves (and their families) a hard-earned livelihood. In many societies, the handling of waste comes with social stigma. Individuals often become waste pickers because it is a low-barrier entry job and largely remain 'out of sight, out of mind' in the societies they service. Waste pickers may belong to religious, social or other minorities, such as (illegal) immigrants, are a financially vulnerable group and are often the urban poor, informal settlement dwellers, pensioners and abandoned children without governmental support, among others. Therefore, informal recyclers are often already socially excluded before starting in this occupation, and are further stigmatised by the profession itself.

Open burning of uncollected waste on a main road in Kibera, Nairobi. PHOTOGRAPH: COSTAS VELIS.

In most cases, waste pickers do not operate within a safe system of work; for example, they do not have access to and/or cannot afford personal protective equipment, or perceive that its use will reduce their productivity. In addition, waste pickers are exposed to all sorts of microbial and chemical hazards and associated injuries. They operate in a low- to no-support system with high-risk conditions and processes. If they are picking in dumpsites, arguably one of the most polluted environments on Earth, their risks are exacerbated. For example, they may be inhaling not just landfill gas, but also the emissions from open uncontrolled burning; they may also be at risk of injury from offloading vehicles or waste 'landslides'. The chances of waste pickers having access to decent health services are slim, which further worsens the damage they may suffer.

Waste pickers are also put in a vulnerable position by other stakeholders or the state itself. Because their activities may be officially illegal, they may be under constant threat of prosecution. In addition, they occupy the base of the secondary materials supply chain: as such, with no access to capital or borrowing, they cannot afford equipment that would add value to the waste, such as transportation means and baling. In that sense, waste pickers are exploitable by those to whom they sell (middlemen, junk shops, aggregators), who often also appropriate most of the value of the materials collected.

Waste pickers' livelihoods depend on fluctuations in the prices of primary materials. Local authorities may see waste pickers' activity as a sign of unwanted backwardness, not compatible with the positive connotations associated with formal solutions. In difficult environments, such as dumpsites or countries with poor law enforcement, waste pickers may need the authorisation of organised crime groups to operate. In some parts of the world, they are subject to political collusion, approached by political parties as clientele, leading to oppression when the governing party is not in power. Waste picking is a hard life, full of danger, instability, low wages and the possibility of exploitation.

Inclusion, organisation and its benefits

In the suite of interventions to improve the situation of waste pickers, many terms are used and debated: 'inclusion', 'legalisation', 'organisation' and 'formalisation'. All of these are underpinned by the intention to make the informal recyclers less vulnerable by empowering them to operate in a more safe, secure way with capabilities to interact with other legal entities and stakeholders. There is an increasing focus on removing societal prejudices and fighting exclusion. However, often IRS workers reject the prospect of being absorbed as workers by formal waste management industry companies.

Increasingly, waste pickers are organising themselves into entities that help them do a better job and secure their income, such as cooperatives and other wider community-based originations (CBOs), as you will read about in some of the programs shared in this book. In some countries and regions, mainly in Latin America, the Caribbean, the Pune region in India and South-East Asia, waste picking is legally recognised as an occupation. In some cases, waste pickers may be contracted by the municipality and become part of the formal economy.

There are simple core ideas that work towards the effective empowerment and inclusion of waste pickers: make their activities legal, include workers in the national registry of occupations and facilitate self-organisation in groups that are legal entities. Other core ideas include providing safe access to waste (e.g. moving away from dumpsites), enhancing the ability of waste pickers to appropriate the value present in the recyclable waste items and ensuring access to health services and education for their children. Importantly, there is a much-needed push for social and financial recognition of the positive role waste pickers play in society.

These ideas have been tested in cases around the world, with some very affirmative outcomes for all stakeholders involved.[1] Yes, there are instances of major

A bale of recycled plastic materials collected by waste pickers in Brazil. PHOTOGRAPH: COSTAS VELIS.

failures, where the purported changes have not addressed core issues such as poverty alleviation or social exclusion, much less the technical aspects of waste management. However, new tools are increasingly available to ensure that interventions are balanced and are addressing all core considerations (e.g. 'InteRa').[2] Ultimately, when operating in an organised fashion where recycling and wider collection services are otherwise unavailable, they constitute a major hope in delivering better waste and resources management on the ground.

Can the IRS help mitigate the global plastic pollution crisis?

Waste pickers are the most obvious and important manifestation of a circular economy practice across the Global South. When it comes to plastic waste, our conservative modelling indicates that waste pickers are responsible for more than half of all plastics recycled worldwide,[3] a remarkable contribution to sustainability both in the form of a viable plastics circular economy and for preventing plastic pollution. A considerable fraction of these salvaged items may have otherwise been dumped in the environment, land and water, or burned. Therefore, one could reasonably hope to see an even greater contribution to plastic pollution prevention by organising and scaling up the IRS activities across the Global South, where the challenges are most prominent.

However, for this to happen, massive and broader changes that go well beyond the IRS are required. In fact, the current global plastics conventional recycling system is suffering from limited overall demand and has been severely negatively affected by recent national and international legal limitations placed on the transboundary trade of plastic waste materials. Therefore, a core enabling step is to move from the current supply-driven plastic recycling economy to one that is driven by demand. For this, end-markets or proven depolymerisation technologies for the flexible plastics (films, multilayer/multimaterial) are still needed.

Currently, waste pickers focus their efforts on the most lucrative materials, such as food-grade polyethylene terephthalate (PET) and high-density polyethylene (HDPE). Other types of plastic are less profitable in general, and some are downright problematic. Fundamentally, the recycling economics currently do not stack up. Simply put, the full cost of safely recovering and converting plastics is not paid in systems supported by unorganised individual waste pickers. In this case, the waste pickers are exploited as low-cost labour, without appropriate health and safety practices in place to ensure their well-being. Positive change means ensuring substantial financial resources are directed to the recycling system. This could include the pricing of recycled plastics and covering costs through product stewardship schemes, such as robust extended producer responsibility frameworks.

Most importantly, can the mass-scale inclusion and organisation of the IRS materialise within a reasonable time scale? This is a necessary precondition to reduce plastic losses to the environment. If systemic changes are prioritised and best practices are more widely shared, there should be hope for a much faster and effective change.

The inclusion and empowerment of waste pickers will be successful when the process provides these front-line workers with a safe, dignified, equitable and secure job and livelihood while ensuring that plastics are sustainably recycled and recovered and plastic pollution is prevented. Despite requiring massive change, this is a sustainable opportunity not to be missed.

References

1   Velis CA, Hardesty BD, Cottom JW, Wilcox C (2022) Enabling the informal recycling sector to prevent plastic pollution and deliver an inclusive circular economy. *Environmental Science & Policy* **138**, 20–25. doi:10.1016/j.envsci.2022.09.008

2   Velis CA, Wilson DC, Rocca O, Smith SR, Mavropoulos A, *et al.* (2012) An analytical framework and tool ('InteRa') for integrating the informal recycling sector into waste and resource management systems in developing countries. *Waste Management & Research* **30**(9), 43–66. doi:10.1177/0734242X12454934

3   Velis CA, Cook E (2021) Mismanagement of plastic waste through open burning with emphasis on the Global South: a systematic review of risks to occupational and public health. *Environmental Science & Technology* **55**(11), 7186–7207. doi:10.1021/acs.est.0c08536

4

Community programs to reduce plastic waste

How to interpret each program

There are many ways to reduce waste and there are many actors (i.e. people) involved. To understand how each program reduces waste by turning it into a commodity, and who is involved, we made a flow chart of the key steps by which a program can reduce waste (orange boxes) and the actors involved at each step (blue boxes).

Waste is the start; whether it is recovered from the environment or from a household, the first step is waste. Waste is collected and **aggregated** by **collectors**, whether they be informal waste pickers, volunteers or contracted collection companies. The aggregated waste is then **sorted** into separate material types and processed, whether that be washed, shredded or baled. The sorting can be completed by informal

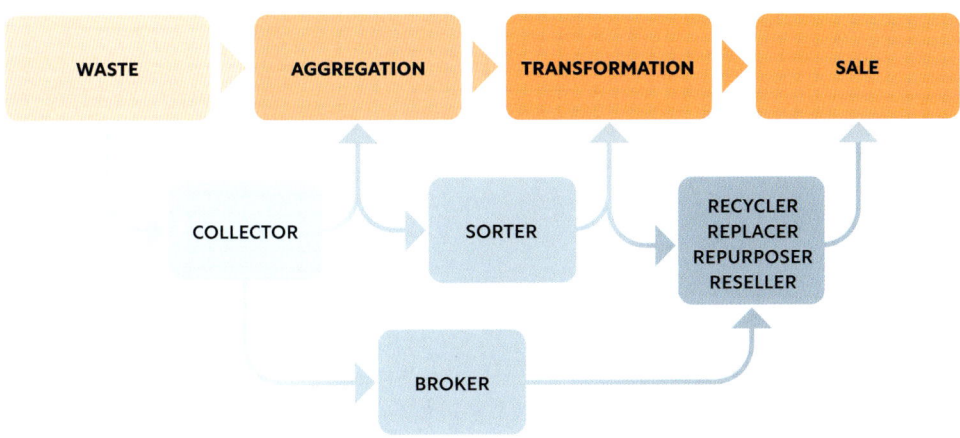

The flow of waste following disposal.

waste **sorters** or by contracted companies. The aggregated and sorted waste is then **transformed** into a new product. Some waste is upcycled, such as plastic bottles into gift boxes; other waste is recycled and extruded, such as plastic polyethylene terephthalate (PET) bottles into recycled PET material. This **transformation** is completed by **recyclers and repurposers**, actors who see waste as a commodity, as something with value that can re-enter the supply chain and not be discarded into landfill. Once the waste is transformed into a new product, that product is then **sold** on to a market. The final actor in this flow is a **broker**. Brokers are not directly involved with the handling of waste material. Rather, brokers enable the transfer/transaction of waste from collectors to recyclers/repurposers. Brokers can enable this transfer/transaction by creating a line of communication between both actors that would otherwise not exist. Brokers help recyclers/repurposers, who create a demand for waste material, connect with collectors, who are suppliers of the waste material.

What components or resources do you need?

 This icon indicates what type and how much waste the program targets. In some cases a program may only target one type of material, such as plastic PET bottles, whereas other programs may target any waste that is discarded into the environment, regardless of whether it is organic, plastic, metal or paper waste.

How long does it take to make operational?

 This icon indicates, first, how long the program took to get up and running from an idea to a fully functioning program. Second, the icon indicates how long the program has been running for. For example, did the program take 6 months to get up and running? Has the program been running for 6 months, 1 year, 10 years? Both times will give you an idea of how long or short some programs can take to implement so you are prepared for the long, or short, journey ahead of you.

How many people will the program serve?

 The final icon indicates how many people the program can serve. For example, does the program provide a waste collection service to a whole municipality? Does the program involve beach clean-ups across an entire country? Does the program operate in multiple countries? The size of the icon will give you an idea of how quickly the program has grown over its time of operation and indicate how scalable the program is. For example, are you

looking to implement a program that will effectively and efficiently manage waste in your community? Or are you looking to implement a program that will fundamentally shift the way waste is valued, or create a new product that can be made and sold around the world?

We think each program included in this book is a unique and fantastic way to reduce waste! We hope you read through each of the stories! May our first page help you decide which ones to read first.

So please peruse the programs we have brought together from 19 organisations across 15 different countries that have successfully reduced waste, turned waste into a valued commodity or otherwise focused on the five Rs (refuse, reduce, reuse, repurpose, recycle) in a novel way, from far up the supply chain all the way to end-of-life solutions.

We hope these programs inspire you!

All Women Recycling

WOMEN EMPLOYMENT | UPCYCLED BOTTLES | SOCIAL ENTERPRISE

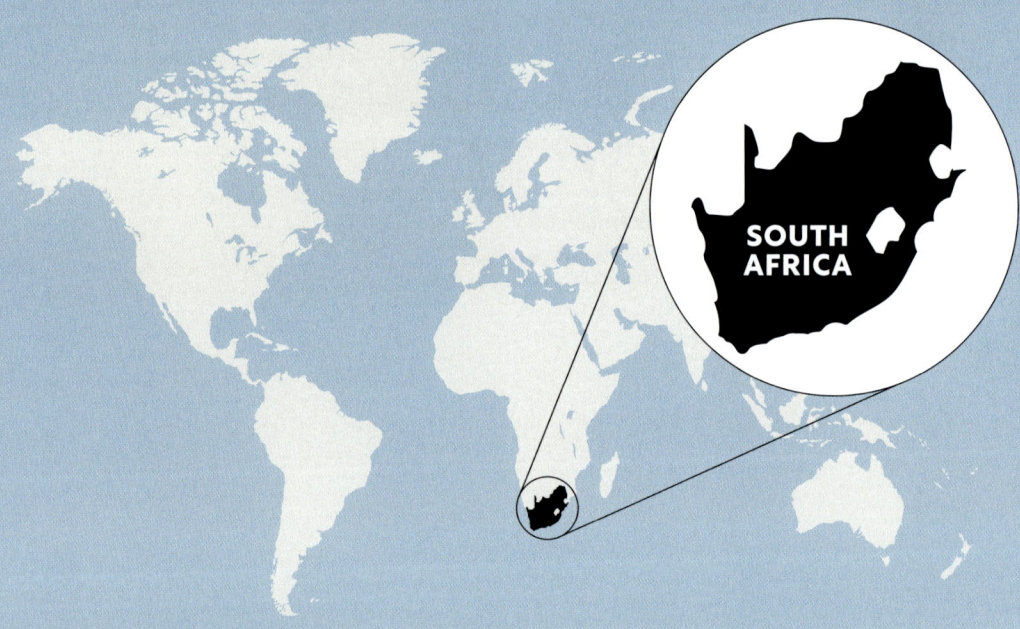

SOUTH
AFRICA

RECYCLER/REPURPOSER

4000 PET bottles
per week

Set up 5 months,
operating since 2010

Multiple
communities

CONTACT

W: https://www.allwomenrecycling.com/   ●   S: @kliketyklik
E: Lynn@allwomenrecycling.com   ●   T: +27 814930297

All Women Recycling founder, Lynn Worsley (far left), and some of the All Women Recycling team.
PHOTOGRAPH: ALL WOMEN RECYCLING.

LYNN WORSELY WAS LIVING IN CAPE TOWN, SOUTH AFRICA, RECENTLY divorced with four children to raise and needing to find a way to make a living. She had to face this problem like an elephant would face a baobab tree: either hit the tree or find a way around it. Always conscious of environmental issues facing South African communities, particularly the over-full landfill situation, Lynn wanted to find a solution that involved reducing the amount of waste entering landfill.

Lynn decided to create a business that was founded on environmentally sustainable principles and empowered women from marginalised backgrounds by providing employment and income, and so All Women Recycling was born. All Women Recycling is a social enterprise that turns plastic polyethylene terephthalate (PET) bottles into marketable products that are sold globally and empowers women locally.

All Women Recycling was founded in 2009 and took 5 months to grow from an idea, through prototyping products, to getting the first retail client order. All Women Recycling is based in Cape Town, South Africa, and has recently expanded its program into Guntur, India. All Women Recycling products are sold in over 30 different countries.

'I chose plastic bottles because of the glut of them all around the world.' – **LYNN WORSLEY, FOUNDER**

How the program works

Training: All Women Recycling puts an emphasis on building the skills of its women employees. The upcycled products require careful craft skills because if the bottles are cut incorrectly, the whole bottle is lost. The women also receive training in a wide range of transferrable skills in production, management and financial administration. This training has already enabled women to move beyond All Women Recycling to further education and employment.

Sourcing material: All Women Recycling buys and collects PET bottles from charities, local schools, street collectors, large partnering recyclers and small recycling companies. By purchasing bottle litter from these groups, All Women Recycling is providing them with an important additional source of income.

Creating products: All Women Recycling upcycles bottles into four unique gift box and container products: kliketyklikbox™, I.PET™, PETTIPOD™ and KEEPET™. Currently 3000 kliketyklikbox™ units are produced per week. These products only use the lower part of the bottle. The upper part of the bottle is returned to informal waste collectors to sell to recyclers, and the bottle tops are donated to a group that supports the provision of wheelchairs to children in need.

Each upcycled product can be tailored to include customer logos or branding and can be decorated with colourful printed serviettes that are imported to the warehouse.

Marketing: All Women Recycling uses media, trade shows, local markets and fairs and radio as marketing channels for their products. All Women Recycling has also partnered with international Fair Trade retailers and the World Wildlife Fund to sell their upcycled products.

Resources

Finances: During the first prototype stages of developing All Women Recycling's first product, the kliketyklikbox™, a friend of Lynn's, who was an executive of a large fashion chain, noticed a kliketyklikbox™ on Lynn's dining room table and asked what it was.

The team at All Women Recycling collect and upcycle plastic bottles into giftwares, such as the kliketyklikbox™. PHOTOGRAPH: ALL WOMEN RECYCLING.

'It's a kliketyklikbox™ ... once a plastic bottle now a gift box to fill with your imagination ... I am going to employ women who need skills and jobs to sell them all over the world.' – **LYNN WORSLEY**

The friend was not sure about the plastic box, but did believe in Lynn's passion and determination, providing Lynn with an investment funding offer (R\$175 000 and 40% of the business) to get All Women Recycling up and running and to produce the first 500 boxes. In 2 years, Lynn had paid her friend back with interest. All Women Recycling is now a self-sustaining business running on the profits from sales.

People: All Women Recycling employs 13 women full-time, some of whom have been with the organisation since the beginning. All Women Recycling also partners with major recycling companies in South Africa, such as Averda, K&C Waste, Polipet and Polyoak, who freely supply PET bottles to the upcycling product line.

Community support: Lynn first advertised for women employees in maintenance courts, but only received one reply. She changed direction and tried advertising by word of mouth. Within 10 days, Lynn had 22 women knocking on her door, in the same position as herself: desperate to find work and earn a living, and to empower their lives. As the business grew, Lynn also grew her relationships with schools, charities and informal waste collectors to purchase plastic bottle pollution and nurture a supportive community.

Equipment: For All Women Recycling to operate, it requires a building where bottles can be turned into upcycled products, an area to package and store orders ready for shipment and tools, such as sewing machines, scissors and glue, to make the upcycled products.

Environmental benefits

All Women Recycling prevents 4000 PET bottles per week, on average, from entering landfill or the environment. This equates to around 17 500 kg of PET per year averted from already over-full landfills in South Africa.

Social benefits

All Women Recycling focuses on employing women, often single mothers, from disadvantaged communities who have been unemployed for two or more years. These women often have very low education and minimal skills training, making it difficult for them to find employment. All Women Recycling focuses on upskilling and educating their employees so women can run their own business and become the engines of growth to get themselves out of poverty.

> 'Not only full-time employment … but we offer daily skills training. So that every day we can honestly say we learn something. Me from them, they from me.' –
> **LYNN WORSLEY**

The demand All Women Recycling has created for PET bottle pollution has also helped reduce poverty and increase education in the broader community. A proportion of All Women Recycling profits goes towards purchasing bottles collected by local schools and informal waste collectors.

All Women Recycling also runs waste-awareness days at schools to teach students to see the value in 'waste to treasure' and entrepreneurship opportunities. Through campaigns with waste collectors, community centres and civil society organisations, disadvantaged communities are also educated in waste and recycling opportunities and their environmental benefits.

Barriers to success

Challenges with raw material: As All Women Recycling grew, it needed more PET bottle pollution to make its products. However, there was strong competition for PET bottle pollution from large recycling companies in South Africa, who sell the material in bulk to foreign markets. All Women Recycling established and supported a network of PET bottle pollution suppliers, particularly from the informal waste sector, to ensure it had a sufficient supply of bottles.

Lack of infrastructure: Within the southern suburbs of Cape Town there is a lack of adequate security and infrastructure. This impairs women's mobility and attendance at work. All Women Recycling had to be careful where it located its production warehouse to ensure women were safe when travelling to and from home.

Scalability and future outlook

Within its South African operations, All Women Recycling plans to reach, educate and empower more women. In the next few years, disabled facilities will be installed

Examples of kliketyklikboxes™ made from recycled bottles.
PHOTOGRAPH: ALL WOMEN RECYCLING.

so women with disabilities can be employed, and there are plans to open a full production and business skills training centre to host a greater capacity of women.

In February 2020, All Women Recycling partnered with Immanuel Bhasker in Guntur, India, to train women to make jewellery for saris from coffee pods in a partnership is called INDIECO. However, the production and growth of INDIECO have been limited by the pandemic.

All Women Recycling is also looking to expand its reach into London over the next year. Lynn views London as a growing green-thinking arena and she wants to develop an eco bus with an England-based environmental company. The bus will travel to schools, teaching environmental sustainability, and sell the All Women Recycling products. A new luxury product, made from 10 bottles per product, is also being prototyped.

Advice from the founder

Leverage existing sales networks: Grassroots sales structures, such as local markets, could be leveraged to increase market penetration faster and with fewer resources than setting up your own distribution points.

Make it different: Market your product as unique with a compelling story behind it to attract attention and to achieve customer satisfaction. Showcase your product as a corporate gift, send samples to potential customers and cross-sell through bundles with complementary products.

Seek out business mentorship: Business development service providers started to support All Women Recycling. This support consisted of advice on management and strategy, mentoring the staff on quality control and pricing and actively profiling All Women Recycling through various media channels.

Create an atmosphere of trust and confidence: Encourage your employees and build on trust and confidence. This promotes a sense of responsibility and an encouraging working atmosphere. If employees work together as a team and towards the same goal, better working results can be achieved. Due to the good work climate, the enterprise recruited a lot of its workforce through word of mouth as current employees referred their friends and family.

Bureo Inc.

REDUCING OCEAN POLLUTION | REPURPOSING WASTE | NEW PRODUCTS

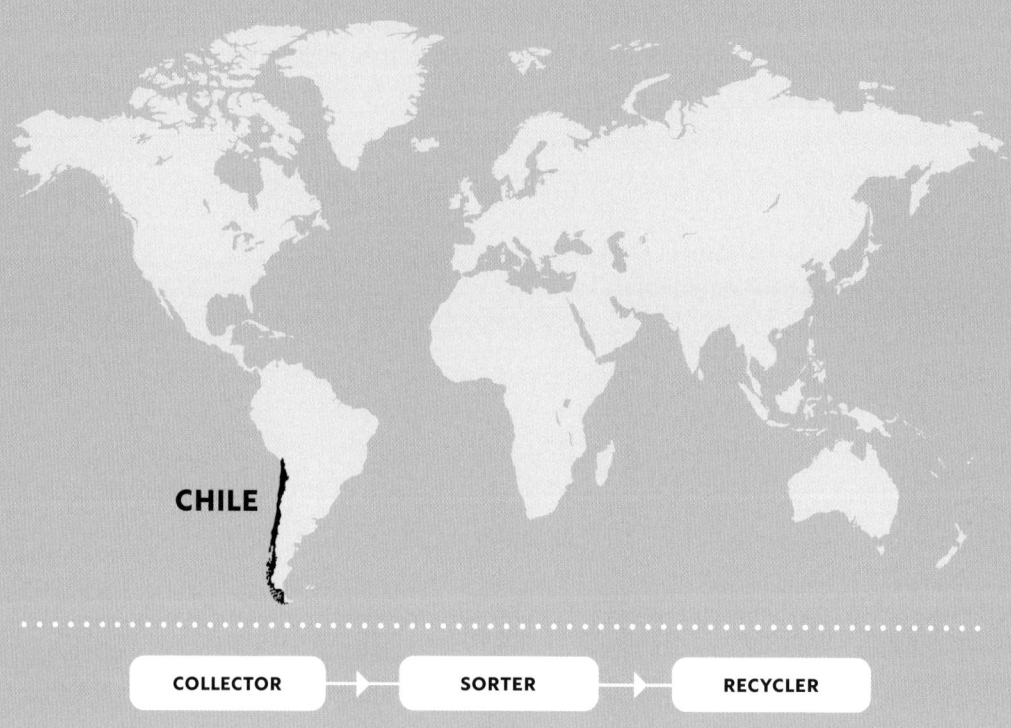

CHILE

| COLLECTOR | → | SORTER | → | RECYCLER |

Fishing nets: Nylon
50–100 tonnes
per month; HDPE
5–15 tonnes per month

Set up 6 months–1
year; operating since
2013

Multiple countries

CONTACT
W: bureo.co/   •   W: bureo.co/blogs/the-rip
S: @BureoInc.; #NetPlus; #UntangleOurOceans   •   E: info@bureo.co
L: https://www.linkedin.com/company/bureo-skateboards/

BUREO (BOO-RAY-OH) STARTED IN LATE 2013, WHEN DAVID STOVER AND Ben Kneppers wanted to combine their respective skill sets as an environmental life cycle assessment consultant and a financial consultant to reduce ocean plastic pollution. After joining forces with a design engineer, Kevin Ahearn, and a few months of research, they cofounded Bureo Inc., a company that transforms ocean plastic pollution into high-value products. A portion of the profits from these products then funds the recovery of more plastic pollution and additional community projects focused on restoring the marine environment.

David, Ben and Kevin spent their childhood surfing and skateboarding, so naturally the first product they trialled making from ocean plastic was a plastic cruiser skateboard deck.

To produce high-value products from ocean plastic pollution, Bureo needed to source plastic pollution that was highly recyclable and durable. Continuing their research into different plastic pollution materials and sources, David, Ben and Kevin discovered that nylon fishing nets were the ingredient they needed. As ideas and product development were forming, Ben was working as a sustainability consultant in Chile. He noticed that fishing nets were, unfortunately, an abundant source of ocean pollution and were causing many problems for the local fishing communities.

Ben connected with local recyclers and fishing industries to launch Bureo's vision of transforming discarded fishing nets into plastic skateboard decks and started Bureo's mission to provide every fisher with an environmentally friendly disposal option for their fishing nets. By early 2014, Bureo had produced their first skateboard deck made wholly from recycled fishing nets.

Since the first skateboard deck was produced, Bureo's mission and operations have grown, and their product line has expanded. Bureo now produces and sells

Christopher 'Caco' Clemo (Chile program lead; left) and Bureo Inc. cofounder Ben Kneppers (right) unloading collected fishing nets at the Bureo warehouse in Chile.
PHOTOGRAPH: ALFRED WESTERMEYER.

Cleaned and packed bales of fishing nets for recycling at the Bureo warehouse in Chile. PHOTOGRAPH: ALFRED WESTERMEYER.

its NetPlus® material made from recycled fishing nets to replace virgin plastic materials in the manufacturing of many products. The NetPlus® material is used in a wide range of high-value products, such as office chairs, bicycles and clothing. Bureo's headquarters are based in Ventura, California, and Bureo has expanded its fishing net collection and recycling operations to Argentina and Peru.

How the program works

Bureo focuses on eliminating one of the most harmful forms of plastic pollution, fishing nets, from entering the ocean. Engaging with local fishers, Bureo recognised that, in most cases, fishers were not the problem, but rather the lack of infrastructure available for fishers to appropriately discard their unusable fishing nets.

Bureo's 100% recycled fishing net material known as NetPlus®. PHOTOGRAPH: ALFRED WESTERMEYER.

Engaging with local communities: The success of Bureo's program is due, in part, to their involvement with local fishing communities. Bureo has engaged with heads of fishery organisations, not-for-profit organisations and fishing communities to form partnerships to retrieve as many discarded fishing nets from the ocean and coastlines of South America as possible.

To get partners involved, Bureo held meetings explaining the direct benefits and incentives for returning fishing nets to Bureo's recycling program, demonstrating how the nets are recycled into a range of products and presenting the final product line. This process has enabled Bureo's program to grow rapidly and allowed multiple fishing communities and fishery organisations in many countries to participate in the fishing net recycling program.

Partnering with brands: Bureo's NetPlus® material has been marketed to a range of clothing, accessories, office equipment and sporting equipment manufacturers. Profits from the sale of NetPlus® finance the continuation of Bureo's fishing net collection and recycling program. The recycled fishing net material can be found in Patagonia hats, Jenga® Ocean™ board games, Costa sunglasses, Humanscale office chairs, Trek bicycles and Carver skateboards.

Resources

Finances: The beginning of Bureo's program was funded through three start-up grants from the Chilean Government Start-Up Chile Program, Northeastern University's IDEA Ventura Accelerator Program and the New England Aquarium's Marine Conservation Action Fund. These three grants provided Bureo with sufficient funds to recover the first fishing nets and produce the first recycled net skateboard decks.

The first full production run of skateboards was funded via Kickstarter. Following this, Bureo received its first financial investment from an outside party, Patagonia's Tin Shed Ventures Fund, which allowed Bureo to upscale its fishing net collection and recycling program to multiple countries in South America. The expansion of Bureo's program throughout Chile and Peru was supported financially by the Chilean Government and the US State Department. This allowed Bureo to source higher volumes of fishing nets and meet the growing demand for their NetPlus® material.

People: Bureo is a member of the Global Ghost Gear Initiative, which promotes and supports the reduction of lost, discarded or abandoned fishing nets around the world. Bureo also partners with Surfrider Foundation, Ocean Conservancy and the 5 Gyres organisation to promote, improve and expand its fishing net recycling programs.

Environmental benefits

Bureo has collected and recycled almost 2000 tonnes of discarded fishing nets. Without Bureo's program and service, these nets would otherwise be discarded in landfill or abandoned in the environment. Bureo's net recovery program includes incentives or compensation for fishers to dispose of their nets through Bureo's recycling program rather than elsewhere. Fishers and other fishing net collectors receive money for every kilogram of net delivered to Bureo's program. Bureo receives nets from over 60 fisheries in South America.

Entering NetPlus® material into the supply chain has decreased manufacturer demand for virgin plastic material. Purchasing NetPlus® has allowed manufacturing companies to shift towards a circular economy system because they are supporting sustainably sourced materials for their products, as well as reducing the negative impacts plastic pollution and virgin plastic manufacturing have on the environment.

Social benefits

Supporting the local community: Bureo uses funds from the sale of fishing nets to work with local environmental non-profit organisations to implement community projects that support artisanal fishing communities that are negatively affected by fishing net waste in Peru, Chile and Argentina. To date, Bureo has supported over 15 community projects, ranging from primary school environmental education projects to community solar power systems and composting systems.

Bureo employs local community members to collect, sort and clean recovered fishing nets at its facilities in each participating country. This employment provides an additional source of income for fishers who participate in Bureo's program.

Discarded fishing net on the shores of an artisan fishing community in Chile.
PHOTOGRAPH: ALFRED WESTERMEYER.

Barriers to success

Changing government legislation: As governments around the world continue to implement solutions to reduce the effects of plastic pollution, the changes in government legislation regarding the transportation of plastics have constrained Bureo's operations to grow. Each new country that Bureo looks to expand into requires a lengthy permit process to allow the transportation and processing of fishing nets. The long permit-processing time restricts how quickly Bureo can set up its program in a new country and start recycling a larger quantity of fishing nets.

Low prices for virgin plastic: As with all plastic recyclers, competing with low market prices for virgin plastic material is challenging for Bureo. The low market price for virgin material can make it difficult for Bureo to compete on price with their NetPlus® material while maintaining fair compensation to all workers involved in the fishing net recycling program and supply chain. Bureo is looking for ways to market its NetPlus® material so buyers will consider paying a higher price for NetPlus® over the cheaper virgin material.

Scalability and future outlook

Bureo is steadily growing its program and the quantity of NetPlus® material supplied to the market. By tracking every fishing net that it has collected and recycled, Bureo is able to complete a life cycle assessment of its NetPlus® material and accurately determine the environmental performance of its program. This assessment will enable Bureo to identify where improvements can be made in its system and feel confident that it is providing a socially and environmentally sound product to its buyers. Bureo's future goal is to continually adjust and improve its operations to achieve an entire circular economy system.

BVRio

WASTE ACCOUNTABILITY | GLOBAL REACH | CIRCULAR ECONOMY

BRAZIL

COLLECTOR → SORTER → RECYCLER/REPURPOSER → BROKER

Up to
500 000 tonnes per
year (predominantly
plastic and glass)

International
operation since 2012

Global

CONTACT
W: https://www.bvrio.org/ ● W: https://www.circularactionhub.org/
S: @BVRio ● E: info@bvrio.org

BVRIO IS A NOT-FOR-PROFIT ORGANISATION THAT STARTED IN BRAZIL IN 2011. Its mission is to design and promote innovative market-based solutions for the benefit of the economy, the environment and the people.

BVRio recognised that there was a low level of investment in waste management infrastructure and programs compared with the magnitude of the pollution and waste problem. There is a huge amount of waste, particularly plastic waste, accumulating in the environment and insufficient measures in place to manage it. In addition, in many countries around the world, there are low collection and recycling rates for many post-consumer materials.

BVRio developed an innovative market mechanism to remove and prevent further leakage of waste into the environment, contribute to a circular economy and improve the livelihoods of local and vulnerable communities involved in waste collection activities.

In 2010, the Brazilian Government enacted a new solid waste legislation that introduced 'extended producer responsibility' to the consumer goods industry. This legislation required that companies contribute to the 'reverse logistics' of products they place on the market to ensure that the products' post-consumption materials are not discarded into the environment, but are instead returned to the economy via recycling.

In response to this legislation, the concept of a performance-based credit system was formed and, in 2013, BVRio piloted a project called Reverse Logistics Credits in Rio de Janeiro. The Reverse Logistics Credits system adapted the concept of carbon credits to the waste management sector. Credits are purchased by producers and importers who need to comply with the solid waste legislation. Each credit funds

An example of a common waste product for coastal countries: ghost fishing gear. PHOTOGRAPH: CIRCULAR ACTION HUB.

(and certifies) the responsible disposal of a certain amount of waste. This responsible disposal of waste involved 100 waste picker cooperatives, known as *catadores*, representing more than 3000 waste pickers, who collected and disposed of over 145 000 tonnes of waste, predominantly glass and plastic, per year.

BVRio engaged with consumer goods companies to purchase the credits relating to the different packaging materials the companies put on the market. A component of the pilot project involved developing an online system to manage and support the waste picker cooperatives to document, account and trace every reverse logistics credit to waste recovered by the cooperatives. The pilot project ran for 2 years, with an average credit worth around US$40.

Due to the success of the pilot project and online management system, BVRio upscaled the program to an international version and created the Circular Action Hub and Circular Credits Mechanism.

How the program works

The Circular Action Hub is an online platform and directory that connects local, scalable waste management projects with companies and investors willing to support, accelerate and strengthen a more effective and socially responsible circular economy. Local waste management projects register their program free of charge and select between the two types of support BVRio provides: direct support and credit purchase. Direct support comprises grants and sponsorship prepayment, or impact investment to the registered project (providing both environmental and financial gains). Credit purchase comprises performance-based payments, where a project delivers a certain amount of circular credits to the interested buyer.

Investors and companies can then search the Hub for projects according to the materials they recover, the location of the project, the volume of material the project recovers or the type of activity the project provides (e.g. collect, sort, recycle, reduce, educate, reuse or some combination thereof). Investors and companies can then fund projects via direct support or by purchasing circular credits.

The Circular Credits Mechanism is a market-based system, like Reverse Logistics Credits, that funds projects that increase waste recovery, sorting and recycling rates, improve the livelihoods of project participants and create and strengthen the circular economy. Similar to purchasing carbon credits to offset a buyer's carbon footprint, circular credits allow buyers to offset their waste footprint that cannot be reduced through internal actions alone; 1 circular credit equates to 1 US ton (0.907 tonnes) of recyclable material that is recovered and adequately disposed of by

BVRio is hoping to see fewer images like this as we move towards a cleaner future.
PHOTOGRAPH: CIRCULAR ACTION HUB.

one of the Hub's participating projects. Circular credits are only eligible for plastic, paper, glass and metal waste recovery initiatives.

Resources

Finances: BVRio is a not-for-profit organisation and, as such, relies on grants, donations and sponsors to finance the beginning of a new project. The pilot Reverse Logistics Credits system was funded by a grant from the Oak Foundation and the Climate Work Foundation. The expansion of the pilot program to an international version, the Circular Action Hub, was funded by corporate member partners and a Partnership for Growth grant. The Circular Credits Mechanism is a self-sustainable system whereby purchased credits finance the commencement and continuation of a variety of waste management programs and activities.

People: BVRio's headquarters are based in Rio de Janeiro, Brazil. To accommodate the expansion and global reach of BVRio's programs, office branches have been established in the UK, Netherlands, China, Ghana and Indonesia. BVRio employs 30 staff.

The Circular Action Hub has a dedicated team of waste management specialists who review and approve projects to be included in the circular credits program. The Hub also has an advisory group, which supports the governance, cooperation and development of the Circular Credits Mechanism and related activities. The advisory group consists of various stakeholders who represent all actors involved in the program. The aim of the group is to guide the governance of a 'learning-by-doing' concept, so that members can decide on what lessons, experiences and approaches from projects and partners around the world can be integrated into the program and foster positive outcomes for all.

Technology: An online platform based on blockchain technology is used to deliver and manage the features of the Circular Action Hub to all its partners. The platform's operations are supported by a team of information technology workers who continually update features for users. The online system allows BVRio to record, manage and track the number of credits issued, transferred and redeemed, thereby providing transparency and traceability to all partners, and ultimately avoiding the double counting of credits.

'True innovation is often very disruptive as it challenges widely accepted assumptions.' – **BVRIO**

Environmental benefits

The Circular Action Hub program hosts over 100 projects, with the capability of recovering and recycling over 500 000 tonnes of material per year. The use of circular credits has enabled companies and investors to engage with and fund local waste management providers in different parts of the world where waste pollution is more prevalent, such as coastal and riverine environments.

Social benefits

Improving the livelihoods of waste pickers: The Circular Action Hub and circular credits system has improved the livelihoods and working conditions for all workers who participate in projects registered on the Hub. The Hub improves livelihoods by ensuring all registered projects adhere to minimum social and environmental safeguards set by BVRio. The Circular Credits Mechanism also provides participating waste pickers and collectors with an additional source of revenue other than the profits they receive from selling recovered material to recyclers. For example, the initial Reverse Logistics Credit program increased the income of waste picker cooperatives by 18–26%.

Establishing polluter-pay regulations: In countries with polluter-pay regulations (such as extended producer responsibility policies), the Circular Credits Mechanism could be recognised as a way companies can comply with the regulations. In countries where polluter-pay regulations do not yet exist, circular credits can be used as a tool for companies and investors to positively contribute to social and environmental initiatives (i.e. corporate social responsibility).

Barriers to success

Lack of trust from partners: Initially, a main constraint on the success of the Hub was a lack of trust from potential partners to participate in the early stages of BVRio's credit systems. To encourage trust, BVRio strengthened cooperation and information exchange between all actors within the program (corporate members, local project developers, local governments and civil society) by involving a wide range of actors in its institutional partners and advisory groups for the Circular Action Hub.

Before the Hub was launched, BVRio called for the plastic recycling community to nominate potential waste management projects it could fund. BVRio received more than 120 applications from over 30 countries. This early enthusiasm from the 'supply' side of the credit system made it easier for BVRio to attract investors, the 'buyer' side of the credit system, to become partners in the program. BVRio believes in developing a market first, then adjusting the market mechanism to suit the market needs.

Traceability of recovered materials: BVRio wants to ensure that the funds from every purchase of circular credits can be traced to finance the recovery of material by waste pickers, who then sell the material to recyclers, who then return the recycled material back to the consumer goods industry, which then returns the recycled material back into the market. An online platform and blockchain technology can be used to overcome this traceability and accountability challenge.

Program variability: Due to the global scope of the Hub, accommodating differences in program types, scope and local contexts of waste management around the world remains a continual challenge. BVRio believes that 'one size does not fit

Waste collectors of the Myanmar Recycling Project.
PHOTOGRAPH: CIRCULAR ACTION HUB.

all', and incorporates the 'learning-by-doing' concept into its project development and evolution. This ensures that BVRio is always adapting and can support the variation in programs that join the Hub.

Scalability and future outlook

The Circular Action Hub and the Circular Credits Mechanism were developed to have global reach. The Hub enables buyers and sellers from all regions of the world to discover and engage with one another so that the recovery and appropriate disposal of waste can occur anywhere in the world, but particularly in areas where waste is an abundant and pressing problem.

The concept of the Circular Action Hub has evolved over time through many iterations. Over this time, BVRio has incorporated 'lessons learned' into the program to address challenges that arise in the recovery and recycling of materials in the most effective and inclusive way.

BVRio has since expanded its initial Reverse Logistics Credits pilot program and now manages multiple programs that promote sustainability and a socially responsible circular economy.

A collection team recovering recyclable plastic from a polluted river. PHOTOGRAPH: CIRCULAR ACTION HUB.

EcoFaxina (Instituto EcoFaxina)

WASTE MANAGEMENT | AWARENESS CAMPAIGNS | REDUCING
OCEAN POLLUTION

BRAZIL

| COLLECTOR | → | ASPIRING SORTER | → | RECYCLER |

66 777 kg marine
debris

Founded in 2008

Multiple
municipalities

CONTACT

W: www.institutoecofaxina.org.br • S: @ecofaxina
E: contato@institutoecofaxina.org.br • T: +55 13 3301 2391

THE INSTITUTO ECOFAXINA (ENGLISH TRANSLATION: ECO CLEANING Institute; hereafter EcoFaxina) is a not-for-profit organisation based in the city of Santos in the São Paulo municipality in the south-east region of Brazil. Founded in 2008, EcoFaxina has coordinated clean-up projects and the promotion of public policies to reduce the volume of plastic pollution entering the marine and estuarine environments in the Santos and São Vicente Estuarine System and along the São Paulo coast. EcoFaxina started with William Rodriguez Schepis moving to Santos in 2006 to study marine biology.

Living in Santos, William noticed that tractors and trucks would clean the local beaches of rubbish each morning so that beachgoers could experience waste-free beaches. William recognised that the rubbish washing up on the local beaches was predominantly coming from the stilted favelas built in the nearby mangrove environment.

When William tried to get information about the favelas and the vast amounts of waste, he found there was none and that no one was organising for this source of waste to be reduced. This spurred him to create EcoFaxina, to clean the local environment of waste and improve waste management within the Santos and São Vicente Estuarine System.

EcoFaxina's primary project is the Environmental Waste Collection System, an initiative to restore degraded mangrove ecosystems and install infrastructure,

Rubbish accumulating around favelas with no waste management. PHOTOGRAPH: INSTITUTO ECOFAXINA.

called 'eco-barriers', to capture litter floating in waterways and improve waste management within Santos. In addition to the waste collection system, EcoFaxina runs voluntary clean-up events with local students and communities.

How the program works

Approximately 50 000 people live across the 14 favelas that occupy large areas of mangroves in the Santos and São Vicente Estuarine System. The informal construction of favelas in the mangroves has resulted in these areas having no waste management or collection systems in place. This has led to many people in the favelas discarding their waste into the nearby mangroves, degrading these environments and causing large quantities of rubbish to occupy the estuarine and coastal environments around the São Paulo municipality.

In 2009, EcoFaxina proposed a partnership with Santos municipality to implement the Environmental Waste Collection System, a three-pronged approach to mobilise favela residents to: (1) manage solid waste through the construction, installation and operation of eco-barriers; (2) restore the degraded mangrove ecosystems; and (3) provide training and environmental education to favela residents and other community members on the benefits of recycling and correctly disposing of waste.

In the near future, all the Environmental Waste Collection System operations will be based out of EcoFaxina's waste management and education facility.

Eco-barriers: Eco-barriers are floating booms that span the width of a waterway, with the ends of the boom fixed to the opposing banks of a waterway. The barriers contain the dispersion of floating solid waste travelling down the rivers and canals of the Santos and São Vicente Estuarine System. This prevents the waste from entering the marine system, where its recovery becomes much harder. The contained waste is collected by favela residents employed by EcoFaxina using barges and boats and is delivered to EcoFaxina's waste management facility, where it is sorted and disposed of appropriately.

In 2016, EcoFaxina received approval for the installation of nine eco-barriers. In the future, EcoFaxina hopes to install more barriers in additional waterways around the estuarine system.

Mangrove restoration program: Mangrove forests around the Santos and São Vicente Estuarine System have become heavily degraded in places where stilted favelas have been built on top of and within the mangroves. Restoring these degraded mangrove forests will restore adequate hydrology to the revegetated ecosystem, with positive effects on biodiversity and the dispersion mechanisms of the ecosystem, protect the shoreline against erosion and help with carbon capture.

Restoring mangroves will involve deconstruction of the existing favelas and providing adequate rehousing for the families currently living there. Once the favelas are deconstructed, the degraded mangrove area will be cleaned of waste and revegetated with locally sourced seedlings by EcoFaxina's large volunteer workforce, biological science students from a nearby university and favela residents employed by EcoFaxina.

The restoration program aims to maintain the natural hydrodynamic flows of the ecosystem, monitor the biodiversity of the area throughout the program and prioritise the restoration of areas where there is a high probability of seedling survival.

Waste management facility: In the near future, the Environmental Waste Collection System will operate out of an environmental education and waste management facility. This facility will have equipment for the delivery and sorting of waste collected from the eco-barriers and favela communities.

Staff employed at the facility will be predominantly sourced from surrounding favela communities. Staff will also be educators and multipliers of environmental awareness within their community, encouraging other favela residents to separate their waste and dispose of it correctly.

Waste collection services currently do not operate in stilted favelas because waste collection in these communities can only occur via the water. EcoFaxina is

A group of EcoFaxina volunteers with the waste they have collected among the mangroves. PHOTOGRAPH: INSTITUTO ECOFAXINA.

Some of the many volunteers cleaning up litter washed up along a riverbank. PHOTOGRAPH: INSTITUTO ECOFAXINA.

currently trialling water-based waste collection services for those favelas that will not be dismantled for mangrove restoration. Floating decks will be located at the end of the main alleys of the favelas and will be points where, twice a week, waste collection vessels will moor to collect waste and recyclable materials. The collected material will then be delivered to the waste management facility.

The waste management facility will also have a voluntary drop-off point where members of the local community can dispose of any recyclable materials. The waste recovered from the eco-barriers, waste collection service, drop-off points and clean-up events will be disposed of appropriately in landfill and any recyclable materials will be sold.

The education and waste management facility will also be a venue to host environmental education workshops for local schools, universities, businesses and communities. The facility will also have an interactive interpretive walk in the nearby mangrove forests for the public to learn about the value of mangrove ecosystems to the community and environment.

Voluntary actions: The Instituto EcoFaxina informs, educates and inspires people to speak and act on behalf of the oceans, emphasising the importance of preserving rivers, mangroves and estuaries to combat marine pollution. EcoFaxina's volunteer activities have been developed to both clean natural environments and draw society's attention to the severe marine pollution and mangrove deforestation in the estuary.

Volunteers remove as much solid waste as possible from coastal ecosystems in the region, especially plastic. These actions also increase environmental awareness and are a research tool providing new perspectives on the problem for those taking part in or following the work.

In addition to the Environmental Waste Collection System, EcoFaxina organises workshops, lectures and corporate volunteering actions that are funded by participating companies or by donors who support EcoFaxina's programs.

Resources

Finances: To fund the Environmental Waste Collection System program and the voluntary clean-up initiatives, EcoFaxina partnered with local government bodies to financially support the program's development. EcoFaxina also received investments from private sector companies that have a shared responsibility in the plastic packaging life cycle and/or work in the marine port and backport facilities within the estuarine system. Finally, EcoFaxina is supported financially by its members and by the sale of its merchandise, such as eco-bags and t-shirts.

People: The EcoFaxina team consists of an executive board of seven members and an administrative council of 11 members. All board and council members work on a voluntary basis without remuneration. The success of clean-up events relies on voluntary participation. Many of the volunteers are students and graduates of nearby universities and schools.

The Environmental Waste Collection System requires the support of residents in the favelas. Favela residents will be directly employed to collect, transport and sort waste materials from the favela communities that have previously not been offered a waste collection service. The waste collection staff will also become recycling and waste disposal educators within their own communities.

EcoFaxina also collaborates with several research organisations to run environmental monitoring programs and to publish the results of these programs in scientific and government publications. EcoFaxina has partnered with the University of São Paulo, the Santa Cecilia University, the Institute of Energy and Nuclear Research and the Paulista State University.

Environmental benefits

To date, EcoFaxina has run 143 voluntary clean-up events that have involved more than 3000 volunteers. The volunteer events have removed around 66 777 kg of waste from the marine, coastal and estuarine environments. Seventy per cent of the recovered waste was plastic. The clean-up events have provided EcoFaxina with the data necessary to make accurate assessments of pollution within the estuarine

EcoFaxina volunteers sorting waste they have collected. PHOTOGRAPH: INSTITUTO ECOFAXINA.

system. These assessments have enabled EcoFaxina to develop strategies and government proposals to combat pollution within the estuary.

Social benefits

Public policies: EcoFaxina maintains a permanent dialogue with local governments and puts forward suggestions and proposals for public policies to mitigate marine pollution and coastal conservation. EcoFaxina's proposals and suggestions have led to the establishment of the following programs by the city of Santos municipal government:

- City Without Garbage Program, which prohibits the disposal of any waste on any beach, footpath, canal, garden or public place within Santos
- Eco-barriers Program, which involves the installation of barriers to contain floating solid waste in the Santos estuary channels
- Paper Bag Dispenser Program, which involves the installation of paper bag dispensers (to replace plastic bag dispensers) in the community for the collection of dog poo.

Community awareness: EcoFaxina started a social media campaign #manguefazadiferenca (#mangrovesmakeadifference) to raise public awareness, appreciation and support for restoring and conserving the mangrove ecosystems around São Paulo.

This campaign focuses on promoting the benefits that healthy mangrove ecosystems provide to the community, such as increased biodiversity, high rates of atmospheric carbon capture and retention, protection of coastal areas against erosion and the silting of estuary channels and providing natural drainage to areas of Santos. The campaign builds community support for EcoFaxina's goal to prevent further construction of favelas in mangrove environments and to establish the Environmental Waste Collection System program.

Barriers to success

Government consultation: EcoFaxina continues to consult with Santos municipal government, particularly the Secretariat of the Environment of Santos, to undertake the recovery and permanent preservation of mangrove ecosystems. Government consultations are a lengthy but important step for EcoFaxina to reach an agreement with the municipal government and receive approval to start the construction of the waste management and education facility for the Environmental Waste Collection System program.

Relocating communities in favelas: Mangrove ecosystems are largely degraded in areas where favelas have been constructed. For these ecosystems to be restored, the favelas need to be deconstructed and the communities relocated to a new area. Furthermore, to prevent the degradation of other mangrove forests, local governments need to prevent the construction of new favelas in other mangrove systems.

EcoFaxina has partnered with COHAB Santista (a company responsible for housing policies in Santos municipality), the regional administration of the north-west zone of Santos and the Civil Defence and Public Prosecution office of the State of São Paulo to ensure current communities and families housed in favelas are provided with suitable alternative rehousing and future housing opportunities.

Scalability and future outlook

EcoFaxina is in the final stages of receiving approval from the Santos municipality to construct the waste management and education facility. EcoFaxina is continuing to raise funds and acquire investors to financially support the construction of this facility. Once approval and funding are finalised, there are no further restrictions to the waste collection program.

EcoFaxina hopes that the success of its eco-barriers, mangrove restoration and waste collection services will inspire other municipalities around Brazil to implement similar programs and reduce the contribution of plastic to ocean pollution by cleaning and restoring degraded areas of mangroves.

GreenHub

NETWORK SOLUTIONS | EMPOWERING WOMEN | RECYCLING

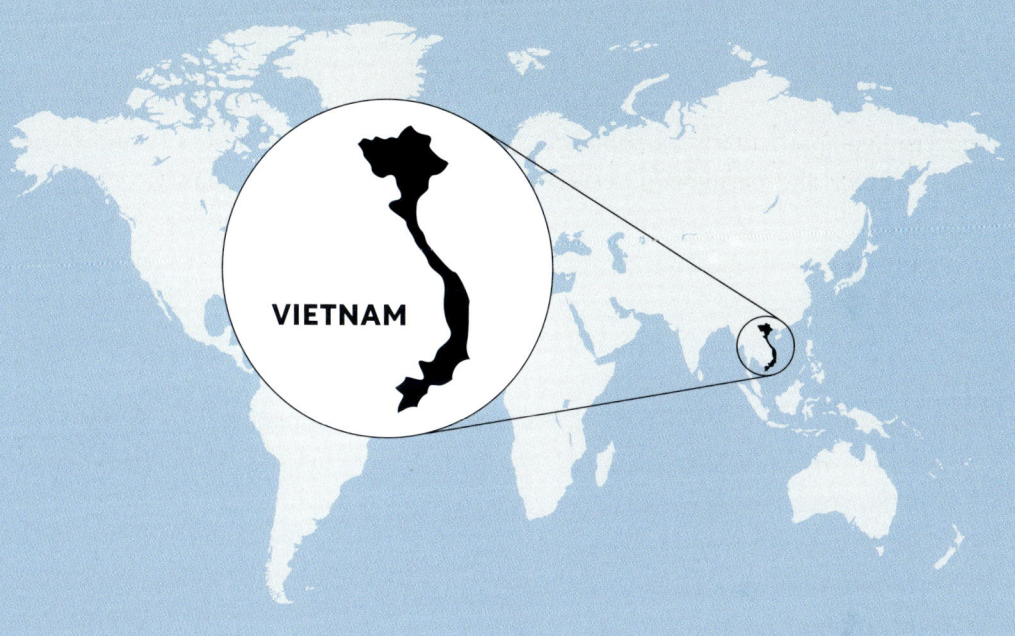

VIETNAM

COLLECTOR → SORTER

Plastic 625 kg per week, organics 2250 kg per week

Set up 7 months, operating since 2018

Multiple municipalities

CONTACT

W: www.greenhub.org.vn • S: @greenhubvn • E: info@greenhub.org.vn

THE CENTRE FOR SUPPORTING GREEN DEVELOPMENT, COMMONLY KNOWN as GreenHub, is a science and technology organisation that creates platforms to connect communities and mobilise resources to practice a green lifestyle, green production and nature conservation.

Over the past two decades, Vietnam has experienced rapid economic growth and urbanisation. Consequently, this has increased solid-waste generation in Vietnam to over 30 million tonnes produced per year; only 10% of that waste is recycled or reused. The remaining waste is discarded in landfill sites, approximately 80% of which are inadequately managed. This has caused around 2500 tonnes of plastic waste to be discharged from Vietnam into the environment per day.

The lack of sufficient waste disposal and the growing quantity of waste entering the environment led GreenHub to establish the Plastic Action Network (PAN), a collaborative program linking local government, businesses, youth groups and women's unions to reduce, reuse and recycle plastic waste and to promote innovative solutions to a circular economy. PAN's goal is to manage the increasing waste production in Vietnam by establishing community-based businesses, particularly for disadvantaged women, that use innovative solutions to reduce, reuse and recycle plastic.

PAN operations commenced in 2018 and are currently piloted in Quang Ninh Province, home to the Ha Long Bay World Heritage Area and where over 87 000 tonnes of waste is produced each year, 16% of which is plastic waste.

To engage tourism enterprises, governments and local communities to preserve and restore the Ha Long Bay World Heritage Area, the International Union for Conservation of Nature (IUCN) and the United States Agency for International Development (USAID), with Vietnamese non-governmental organisations, businesses and governments, formed the Ha Long Cat Ba Alliance. Partnering with this alliance and the women's unions in Quang Ninh Province, PAN was formed by Nguyen Thi Thu Trang, deputy director and cofounder of GreenHub, under financial support from The Coca-Cola Foundation.

In December 2018, PAN held the first training course for the Women's Union in Ha Long, Quang Ninh Province. The training focused on building the knowledge and technical skills of women to start their own initiatives that reduce domestic waste and the use of single-use plastic products.

The PAN program trains disadvantaged women to start their own businesses, partner with waste pickers and improve their livelihoods by turning various plastic wastes into upcycled eco-bricks, sun hats, flower vases and bags. Eco-bricks are plastic beverage bottles that have been tightly filled with unsellable plastic bags to form solid plastic bricks that can be used for small structures, such as garden beds.

Other initiatives turn organic waste into compost or sell waste materials to recyclers. In addition to these initiatives, PAN runs campaigns in the community to raise awareness of the harmful effects of plastic pollution, the benefits of separating waste and considering waste as a resource by 'turning trash into cash'.

How the program works

PAN has five stages of operation: communication, collecting, sorting, reusing and recycling.

Communication: Women's unions encourage households to sort waste and keep plastic straps and single-use plastic bags.

Collecting: Collections occur on weekends, with female waste pickers and members of the women's unions collecting waste directly from households, restaurants and hotels. Households can also drop off their waste at designated community house collection points. This requires PAN to promote awareness campaigns and establish relationships between the women's initiatives and the households and businesses their program services.

The collection services also require forms of transport to make the collection and transportation of waste materials easy. For example, some women use scooters or bicycles, whereas other women requested a garbage truck from their local government for their collection service.

Sorting: Waste is sorted into materials for upcycling and reusing initiatives and for selling to recyclers. The reused waste material requires a space where it can be washed and prepared. These processes often occur in women's households or in alleys.

Reusing: Women who have started their own business initiative after participating in a PAN training course make eco-bricks and handbags from sorted

A GreenHub team member sewing recycled trash into giftware. PHOTOGRAPH: GREENHUB.

and cleaned plastic waste materials. PAN establishes and promotes marketing for upcycled products handmade by local women.

Recycling: Collected plastic waste is sold to recycling agencies that make a range of recycled products. PAN supports the promotion of marketing recyclable materials and selling to recycling agencies.

'We don't think that we've taken part in a 3-year pilot project; instead, we consider it to be a permanent project because environment protection is the lifelong responsibility of every citizen.' – **REPRESENTATIVE OF THE WOMEN'S UNION OF HẠ LONG CITY PARTICIPATING IN A PAN PROGRAM**

Resources

Finances: In August 2018, GreenHub received a 3-year grant funded by The Coca-Cola Foundation to launch the first PAN plastic waste management programs. Before any PAN programs commenced, GreenHub consulted with multiple government departments in Quang Ninh Province (e.g. the departments of Environment and Natural Resources; Construction, Industry and Trade; Education and Training; Science and Technology), Ha Long Bay management board and Viet Long Joint Stock Co.

During these consultations, GreenHub introduced and shared the project goals, explained how to organise the project effectively and informed stakeholders of upcoming activities and how they could support them.

People: For PAN programs to operate, a project team is required to manage communications regarding awareness campaigns and PAN events; coordinate and facilitate resources and training courses; and consult with women's unions, local governments and businesses. PAN brings together a range of organisations with expertise in business, waste management and innovative solutions to collectively work towards reducing waste in Quang Ninh Province.

Environmental benefits

The PAN program has successfully increased community perceptions of the value of waste. Establishing waste collection, repurposing and recycling programs has significantly decreased the amount of waste discarded in the environment and to landfill within the Ha Long region of Quang Ninh Province. Within 2 years, PAN activities have collected over 101 tonnes of waste, 60% of which was plastic waste.

Social benefits

Improving livelihoods: During the 2 years of its operation, PAN has fostered a range of collection initiatives, educated and partnered with over 3700 female waste

A range of baskets produced by GreenHub from waste materials.
PHOTOGRAPH: GREENHUB.

pickers and trained over 175 disadvantaged women in sorting waste, composting and eco-brick production. PAN-supported women initiatives have diverted 54 tonnes of organic waste into compost, saving over VND$24.3 million in waste treatment costs, and created over VND$15.6 million for local women from the collection and selling of 47 tonnes of plastic waste for recycling or reuse.

In addition, the PAN programs have made 450 eco-bricks that have contributed to public works in the community. PAN awareness campaigns have reached 20 000 members of the public and led to over 5600 people participating in project activities.

The project has made a VND$130 000 000 profit from selling waste materials. These profits are used by women's unions for fundraising, charity and environmental protection work within the community.

Education: PAN has a strong education component, particularly focusing on women in the community. Education programs raise participant awareness of environmental issues, the harms of plastic pollution, the value of separating waste streams and considering waste as an economic resource. The PAN training workshops have also built the fundraising, planning and environmental activity capacity of women participants.

Since PAN started, communities in the Ha Long region now have household waste collection services, community-based initiatives that turn waste into a valuable product and initiatives that reduce the volume of valuable waste entering landfill.

Barriers to success

Finances: The grant that kickstarted the PAN program lasted for the first 3 years of operation. However, upon completion of those first 3 years, the program was not yet

The GreenHub team turning plastic bottles into giftware. PHOTOGRAPH: GREENHUB.

fully self-sustainable and required additional funding to continue operations. PAN continues to work on establishing more markets for upcycled product initiatives and recyclable materials to achieve a sustainable system.

Challenges with raw materials: A challenge for PAN operations is finding sources of materials to upcycle. For example, establishing a regular source of plastic strap waste from the construction industry has been difficult. Many construction sites burn or shred their strap waste. GreenHub worked with waste pickers and construction sites to establish a regular strap waste collection service rather than having the straps burned on-site.

The supply of straps from construction sites is seasonal, with less strap waste produced during the rainy season when there are fewer construction works. Waste pickers are exploring storage options for the collected straps to ensure they have enough material to create upcycled products throughout the year, even when strap waste production is lower.

Scalability and future outlook

The PAN program is currently being piloted in the Ha Long region of Quang Ninh Province to identify and resolve any complications and streamline program systems so that PAN can be scaled to cover the entirety of Vietnam.

Life Out of Plastic (L.O.O.P.)

REPURPOSING WASTE | REDUCING OCEAN POLLUTION | AWARENESS CAMPAIGNS

PERU

COLLECTOR → RECYCLER/ REPURPOSER

140 000 tonnes of marine debris, 1.3 million PET bottles

Set up 12 months, operating since 2012

Countrywide

CONTACT

W: http://www.loop.pe/ • W: https://hazla.pe/ • W: http://loop.pe/mundo-loop/

S: @lifeoutofplastic • L: https://www.linkedin.com/company/life-out-of-plastic-sac/

E: comunicaciones@loop.pe

IN 2011, ON VALENTINE'S DAY, IRENE HOFMEIJER WAS REFLECTING ON A recent visit to the Peruvian Amazon where she saw the reality of the harms that mismanaged waste was doing to the environment. Witnessing a place that she cared for become degraded by our waste, she decided to start an initiative that would act on plastic pollution and reduce the damage it caused. Soon after starting, Irene met Nadia Balducci, a marine biologist working with fishing communities in northern Peru, who, like Irene, was confronted by the extent of the effect plastic pollution was having on the environment.

Together, Irene and Nadia launched a social impact company that would not only help the environment, but also shape young leaders, particularly women, and create a community of people pushing forward for change. Life Out of Plastic (L.O.O.P.) was born, an entity that combats plastic pollution through local action, social campaigns and the creation of a product line made from recycled plastics.

'Plastic started as a material that was easy to use and has now become waste in a world heavily turned to consumption.' – INÉS YÁBAR, COMMUNICATIONS OFFICER AT L.O.O.P.

At the end of their first year of operation, L.O.O.P. had successfully launched a product line of tote bags made from recycled plastic waste, offered a corporate

The founders of L.O.O.P. PHOTOGRAPH: L.O.O.P.

volunteering campaign service and established regular plastic pollution awareness campaigns with a national beach clean-up campaign and a consumer-conscious art show.

How the program works

L.O.O.P. has four components to achieve its aim of creating awareness and reducing plastic pollution: sustainable products, creating awareness, citizen movement and transforming companies.

Sustainable products: L.O.O.P. is targeting consumers to reduce their plastic consumption and waste generation by creating and providing a market for products made from recycled polyethylene terephthalate (PET) plastic. L.O.O.P.'s tote bags are targeted to the retail sector to motivate shoppers to replace single-use plastic bags with L.O.O.P.'s reusable tote bags.

A campaign for the sale of tote bags generated publicity and consumer demand, predominantly in Lima, the capital of Peru, by highlighting the benefits of purchasing reusable products made from recycled PET plastic over virgin PET. L.O.O.P. also has a 'conscious gifts' catalogue for corporate merchandise products. The catalogue contains reusable, upcycled products made by conscious entrepreneurs and L.O.O.P. itself. The catalogue is distributed to companies and institutions to offer them alternatives to their traditional merchandise and to support a market of local sustainable products.

Creating awareness: Through school visits, social media and online blogs, L.O.O.P. generates awareness of plastic pollution and how people can participate to minimise their waste and use of single-use plastics. L.O.O.P. seeks to uproot from society the easy, disposable culture of buying, using and throwing away without asking where the product came from or where it ends up.

Citizen movement: L.O.O.P.'s campaigns and activities rally citizens to be active participants in marine conservation and offer concrete solutions for them to live a life without plastic. L.O.O.P.'s largest campaign is *HAZla por tu playa* ('Do it for your beach'). The HAZla campaign mobilises thousands of citizens across Peru to clean their local natural areas during the first week of March.

All litter collected is recorded in a standardised way so that L.O.O.P. can track the impact that their social movement campaigns have on the type and quantity of waste in the environment. L.O.O.P. co-organises the national event with another organisation and formal recycling partner to ensure that the recovered waste is recycled or disposed of correctly. Within Lima, L.O.O.P. also co-organises smaller local campaigns with civil society partners that promote sustainable consumption and waste minimisation.

Conducting litter surveys at a beach. PHOTOGRAPH: L.O.O.P.

'We believe that everyone has a responsibility to tackle plastic pollution regardless of race, gender, background or age. This is what the HAZla movement has allowed.' – **INÉS YÁBAR**

L.O.O.P. believe that cleaning beaches is not the solution to plastic pollution. However, people can be shocked by seeing plastic pollution and picking it up from the beach, even more so when they pick up products they recognise and consume frequently. L.O.O.P.'s beach clean-up campaigns generate an emotional response in people that drives them to make a series of conscious changes to their consumption behaviours.

Transforming companies: L.O.O.P. is leading a campaign that raises awareness about the importance of recycling with two major supermarkets in Peru. The campaign allows people to learn about responsible consumer choices and how to recycle correctly. Better recycling behaviour has the added benefit of assisting formal recyclers, who will have access to separated waste streams that they can recycle easily.

Resources

Finances: The initial starting capital for L.O.O.P. was entirely from Irene and Nadia, who invested their savings into the company. Up until 2016, L.O.O.P.'s activities were primarily funded through its corporate volunteering services and the sale of tote bags made from recycled PET plastic.

The revenue from these market-based activities finances L.O.O.P.'s social initiatives, such as beach clean-ups. Due to the increased competition from imported, lower-cost recycled fabrics, L.O.O.P. discontinued its product line and is now funded through its corporate services. L.O.O.P. has also received small grants from social enterprise competitions to finance its activities.

People: Most L.O.O.P. activities take place in Lima and rely on a network of volunteers to lead beach clean-ups and awareness-raising activities. For some campaigns, L.O.O.P. partners with organisations that work with formal recycling facilities. L.O.O.P. is run by a mix of staff and volunteers based in Lima.

A core team of 12 volunteers mobilises a network of thousands of volunteers for L.O.O.P. campaigns. Core team volunteers are given opportunities to lead projects based on their expertise, earning a small stipend, and are hired as staff on a project-by-project basis. The core team is led by two staff, who were previously volunteers, and supported by two part-time administration and accounting staff members. The founders of L.O.O.P. remain active at an advisory board level and meet with the core team monthly.

Community support: Since 2011, L.O.O.P. has built community support for its company and activities through campaigns and community events. Over the years, L.O.O.P. has run art exhibitions on the topic of plastic pollution, beach clean-up events along the coast of Peru and waste education seminars at schools, particularly in disadvantaged communities where environmental education is uncommon.

'Our communities are the heart of our conservation efforts.' – **INÉS YÁBAR**

Community support has been instrumental in L.O.O.P. building a social movement and an army of volunteers to reduce plastic pollution. Without this support, L.O.O.P.'s efforts to reach the public and private sector would have taken longer.

Equipment: L.O.O.P.'s beach clean-ups require reusable sacks to store the rubbish recovered from the beach, personal protective equipment for volunteers to safely participate in the event and often a highly visible shirt for participants so they are easily identifiable along the beach.

Environmental benefits

L.O.O.P. upcycled 1.3 million PET bottles through its recycled PET product line. Using recycled PET fibre to produce its products meant 76% less greenhouse gas emissions than if virgin polyester fibre had been used. Beach clean-ups run monthly and are led by a group of keen volunteers who mobilise participation from the local community. The clean-ups have removed over 140 000 tonnes of marine debris from the Peruvian coastline.

Data collected from the HAZla campaign was used as key evidence for 15 bills presented to the Peruvian congress to implement laws regulating the use of single-

Microplastics are another consideration in marine debris pollution and are regularly found in beach surveys. PHOTOGRAPH: L.O.O.P.

use plastics in Peru. The bills led to the government implementing Act No. 30884 in 2018, which sets deadlines for the eradication of single use plastics.

Social benefits

To date, over 34 000 people have taken part in L.O.O.P.-organised activities. Approximately 17 000 women have participated in a L.O.O.P. training session or campaign or have volunteered in a L.O.O.P.-organised activity. L.O.O.P. is a 100% women-led enterprise that promotes female leadership and capacity building by implementing a women-only hiring gender bias and women-only campaign coordinators, and striving to involve women-led organisations as partners.

Barriers to success

Shifting waste: L.O.O.P.'s national campaign has assisted in generating social awareness and government action in reducing the use of single-use plastics. Despite this, L.O.O.P. is not noticing a decrease in the quantity of waste discarded on beaches, but rather a shift in the type of waste discarded. L.O.O.P. is now starting to see new types of waste littering beaches.

Instead of seeing single-use plastic bags and straws on the beach, L.O.O.P. is now finding compostable bags and cigarette butts. L.O.O.P. recognises that its campaigns are having an impact, but that there is still more to do with regard to improving citizens' waste behaviours.

Maintaining enthusiasm: Beach clean-up participants can often feel discouraged by their cleaning efforts when more plastic washes up on the beach the following day. L.O.O.P. volunteers can also become discouraged by the slow change

One of L.O.O.P.s campaigns is a national beach clean-up in Peru. PHOTOGRAPH: L.O.O.P.

in communities who continue to discard waste inappropriately or who believe there is a lack of evidence to support actions that prevent the harms of plastic pollution.

Through regular engagement and communication through blogs and social media, L.O.O.P. promotes the actions of its beach clean-up participants and volunteers and continues to lobby governments and the private sector to act to reduce plastic pollution. L.O.O.P. runs capacity-building and leadership programs for its volunteers so that they get the most out of their volunteer experience while making the greatest impact with their actions.

Scalability and future outlook

Although L.O.O.P.'s tote bags made from recycled PET fibre have been discontinued, it does maintain a small product line made from recycled cotton available for one-off orders. Prior to the COVID-19 pandemic, L.O.O.P.'s operations model was self-sustainable. However, due to financial pressures from the pandemic, L.O.O.P. reviewed its model to allow outside investment for new projects and to continue generating major change in Peru's public and private sector. L.O.O.P. is developing new services that not only address the downstream issue of plastic pollution, such as beach clean-ups, but also help community members in their responsible consumer choices upstream and to avoid single-use plastics.

Mamma's Laef Vanuatu

REDUCING WASTE | SANITATION | COMMUNITY-BASED

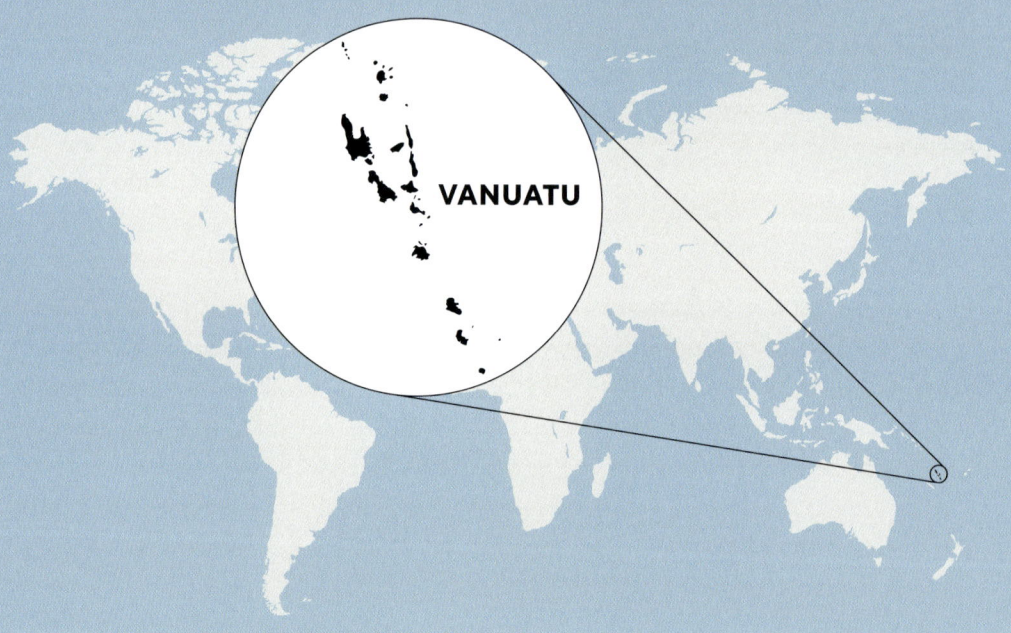

VANUATU

REPLACER

Replacing single-use
hygiene items with
reusable items

Set up 2 years,
operating since 2017

Countrywide

CONTACT
W: https://www.mammaslaef.com/ • S: @MammasLaef
E: hello@mammaslaef.com • T: +678 543 4414

FOR THE PAST 11 YEARS, BELINDA ROSELLI HAS BEEN A REGULAR VISITOR TO Vanuatu for holidays and to procure Vanuatu-produced products for her small social enterprise. In March 2015, Cyclone Pam, the largest Category 5 cyclone to ever be encountered, hit Vanuatu. Returning to Vanuatu soon after the cyclone, Belinda noticed that access to menstruation and hygiene products was heavily restricted for many ni-Vanuatu communities. She asked, 'What do women do in times of disaster for menstruation? How do they access hygiene products when supply chains for single-use products are disrupted?'

Belinda found that many women and girls had to use undignified alternatives, such as rags. In April 2015, Belinda distributed aid packages that contained disposable hygiene products to cyclone-affected communities, but she saw that this was not a long-term solution.

Belinda returned to her home country of New Zealand and started investigating a longer-term solution to the broken supply chain for hygiene products. She researched reusable products and, in October 2015, returned to Vanuatu and ran a series of scoping workshops to determine community support and facilities for using and manufacturing reusable hygiene products. There was interest from some communities, so Belinda partnered with a ni-Vanuatu couple, Mary and Jack Kalsrap, to make 100 reusable menstrual pads and travel around markets selling them.

Due to the huge interest in the products from women around Vanuatu, Belinda worked with Mary and Jack to start their own social enterprise, Mamma's Laef (pronounced life).

Some examples of Mamma's Laef reusable products. PHOTOGRAPH: MAMMA'S LAEF

Mamma's Laef continued as a small community-based and -supported organisation manufacturing and selling reusable hygiene products throughout Vanuatu until in 2018, when it received a Frontiers Innovators grant financed by the Australian Department of Foreign Affairs and Trade. Mamma's Laef was one of 15 groups to receive the grant from 700 applicants. The funding enabled Mamma's Laef to become a formalised, Vanuatu-owned, legal entity and establish a charitable trust based in New Zealand, the Mamma's Laef Charitable Trust. The Trust was set up to help guide the development of Mamma's Laef in Vanuatu.

'Mamma's Laef empowers women by providing sustainable menstrual products, creating social enterprise, and breaking taboos about menstruation.' – MARY KALSRAP, CO-OWNER OF MAMMA'S LAEF

How the program works

Mamma's Laef is an enterprise where local women are given the opportunity to develop their vocational skills and earn an income by manufacturing three main hygiene products: menstrual pads, adult continence aids and baby nappies. The first Mamma's Laef product was a reusable menstrual pad.

The success of the menstrual pad led to women in the community asking Mamma's Laef to design a reusable continence aid for adults, because buying disposable versions was very expensive for communities. Partnering with the

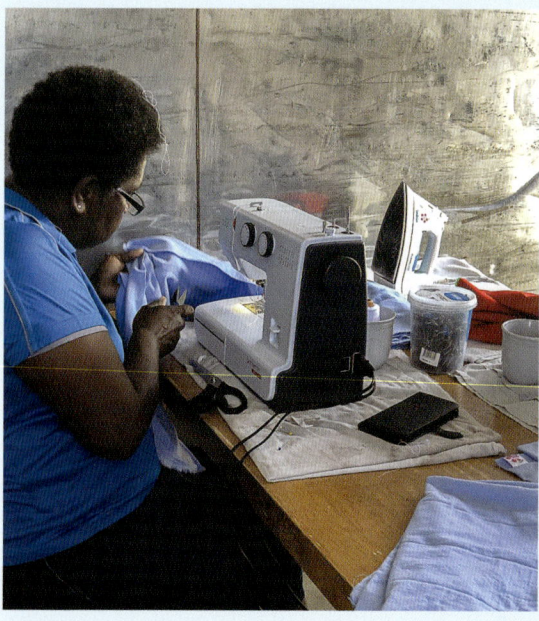

All Mamma's Laef products are produced locally by local people. PHOTOGRAPH: MAMMA'S LAEF.

Canada Fund and Vanuatu Society for People with Disabilities, Mamma's Laef produced adult continence aids for distribution to people living with disabilities or those with some level of incontinence.

Following the success of the adult continence aids, Mamma's Laef revisited a suggestion made to them in 2015 by local tourism industry workers, who had asked Mamma's Laef to design reusable baby nappies. This was suggested because tourism operators were encountering large accumulations of used disposable nappies in the coastal and marine environment. At the time of the suggestion, it was not viable for Mamma's Laef to produce reusable baby nappies. However, in March 2019 the opportunity to produce reusable baby nappies became potentially viable when the Vanuatu Government legislated a ban on disposable nappies.

Reusable nappy pilot study: In 2018, it was estimated that Vanuatu used around 7 million disposable nappies per year, with disposal nappies making up 27% of rubbish in Vanuatu. Mamma's Laef partnered with Bambino Mio, Europe's largest manufacturer of reusable nappies, and Savvy Vanuatu to conduct a pilot study that surveyed the community's support for and use of reusable baby nappies as a solution to the ban on disposables. A full report for this pilot study is available on the Mamma's Laef website.

The pilot study was very successful, with 96% of participants liking the reusable nappies and 85% stating they would purchase them. In addition to the survey, reusable nappy packs were trialled in 60 families for 2 weeks across three peri-urban and rural communities to identify and resolve any barriers to communities using the nappies (e.g. developing methods to hygienically clean the nappies in communities that have limited or no running water).

Mary Kalsrap (left; co-owner of Mamma's Laef) teaching teens about menstrual health.
PHOTOGRAPH: MAMMA'S LAEF.

Education: Non-government organisations operating in Vanuatu are the largest buyers of Mamma's Laef's products. Mamma's Laef also manufactures and distributes products to communities through donations to the charitable trust. When products are given to communities, Mamma's Laef also sends educators from its team to provide information on the use and benefits of the products and teach school groups, women, girls, men and boys in the community about menstrual health.

The awareness program allows residents in the community to be nominated to become champions of the reusable products and to continue to support environmentally friendly initiatives once the educators have left.

Resources

All Mamma's Laef operations, including administration, product design and manufacturing, are based in Mamma's Laef Hub. The Hub is made of two shipping containers located on Mary and Jacks' property in Pango village, just outside of Port Vila, Vanuatu. The Hub has reliable access to electricity and is close to the marine port, minimising domestic transport costs of imported materials. Mamma's Laef also uses a neighbouring shipping container to store imported bulk materials for product manufacturing.

The working space is somewhat congested and expansion is a medium-term plan for Mamma's Laef, with a focus on retaining operations in the village and maintaining the local community connection that is important to the enterprise's ethos.

Raw materials, including waterproof and absorbent materials, are procured from China. Initially, both treadle- and electric-powered machines were used to sew and make products, but as skills and resources have allowed, the treadle-powered machines have been phased out and all the production staff now use electric machines.

Finances: Since the beginning, Mamma's Laef has financed its enterprise from the profits from sales and donations to the trust (e.g. the Hub was funded through a philanthropic donation). Through a variety of partnerships, Mamma's Laef has been able to develop and pilot new products, as well as to distribute its products and awareness program to remote island villages.

Most recently, Mamma's Laef partnered with Engineers Without Borders to trial different nappy-washing methods (e.g. hand-operated washing machines) to develop a cost-effective, hygienic and efficient method for communities, particularly those in remote and rural areas of Vanuatu.

People: The enterprise is owned by Mary and Jack Kalsrap and is supported in partnership via a memorandum of understanding with the New Zealand-based Mamma's Laef Charitable Trust, with trustees Belinda Roselli and Tina Onnes. When Mamma's Laef is operating at full capacity, it employs 10 staff, in addition to several temporary staff who are employed during busy periods.

Community support: Initial scoping workshops for hygiene products were conducted with Mary and her community. Mary and Jack are ni-Vanuatu and are very well connected and respected by many communities around Vanuatu. This connection and respect enabled Mamma's Laef to have open conversations with community members about the different products, run education programs with communities and identify which members would be Community Champions. Community Champions were chosen in each new community to ensure culturally acceptable and effective communication.

During the pilot study, Community Champions selected and supported the families in their community who trialled the reusable diapers. Community Champions also provide ongoing support and information about Mamma's Laef reusable menstrual products.

> 'I am very happy to be using this local product because it helps save a lot of money. I want to thank Mamma's Laef for producing this nappy to help jobless mothers.' – **PILOT STUDY PARTICIPANT, PANGO VILLAGE**

Mamma's Laef reusable products can be used time and time again. PHOTOGRAPH MAMMA'S LAEF.

Environmental benefits

Mamma's Laef menstrual pads have a lifetime of 3–4 years. Hence, a woman can use one reusable menstrual pad rather than 600 disposal pads over the same period of time. This replacement reduces the demand for plastic material used to manufacture the disposable products and reduces the amount of disposable waste generated. Menstrual pads have been distributed to 6500 girls and women in communities across 13 islands of the Vanuatu archipelago.

Mamma's Laef is currently measuring the environmental impact of its reusable baby nappies. Similar to the reusable pads, the use of reusable nappies will reduce the demand for and subsequent waste of plastic material.

Social benefits

Gender equality: The Mamma's Laef menstrual health education awareness program has been shared in 32 schools, and in communities across five islands, teaching more than 4500 women, girls, boys and men. The program promotes gender equality by providing a safe environment in which to learn in a culturally appropriate manner to help normalise conversations and remove community taboos around menstruation.

Sustainable Development Goals: Mamma's Laef programs and products support Vanuatu in achieving 10 of the 17 United Nation's Sustainable Development Goals (SDGs). Mamma's Laef reduces poverty (SDG1), provides decent work and economic growth (SDG8) and promotes industry, innovation and infrastructure (SDG9) by operating a ni-Vanuatu owned enterprise that employs local ni-Vanuatu women and men. It improves the health, well-being (SDG3), clean water and sanitation (SDG6) of communities, and improves the health of life below water (SDG14) by providing reusable hygiene products in place of disposable products.

The organisation also provides quality education (SDG4), improves gender equality (SDG5) and reduces inequalities (SDG10) through its education awareness program, and reduces the community's carbon footprint (SDG13) by reducing the demand for single-use, disposable plastic materials.

'Mamma's Laef products represent a good investment in our country. Vanuatu-made products are really important as they build value in our economy, provide work opportunities for ni-Vanuatu people and to keep the money within our economy.' – **FALVIANNA RORY, SENIOR INDUSTRY OFFICER (LARGE SCALE MANUFACTURING) DEPARTMENT OF INDUSTRY, VANUATU**

Barriers to success

Mixed community support: When the Vanuatu Government enacted legislation that banned disposable nappies, some urban communities disapproved. In addition, some childcare centres were unsupportive of reusable nappies because they saw it as extra work, having to clean the nappies. The tourism industry was also concerned that the unavailability of disposable nappies would make Vanuatu an unattractive destination for families with children.

As the government has yet to enact the legislation, concerns have been allayed and Mamma's Laef continues to actively highlight the huge environmental and social benefits that the ban on disposable nappies and the use of reusable nappies can bring to Vanuatu.

Sourcing raw materials: The border closures brought on by the COVID-19 pandemic and another Category 5 cyclone, Cyclone Harold, which hit Vanuatu in April 2020, have disrupted supply chains of materials and hygiene products imported into Vanuatu. The result has been an increase in the shipping costs and shipping times (double) for importing the raw materials needed to make Mamma's Laef products.

To overcome supply chain disruptions, Mamma's Laef received a small amount of relief funding from the Australian Department of Foreign Affairs and Trade so it can afford the increased shipping fees and continue making and distributing its

Education is a crucial part of the impact Mamma's Laef has on local communities. PHOTOGRAPH: MAMMA'S LAEF.

much-needed products to communities in Vanuatu. Mamma's Laef is also looking to diversify its supply chain of raw materials in case one supplier cannot fulfil the order.

Retail price of products: For many women in Vanuatu, the upfront costs of Mamma's Laef's reusable nappies are expensive compared with disposables alternatives. Through their community education program, Mamma's Laef explained the long-term savings of using reusable nappies and encouraged women, families and villages to create a fund for new mothers in their community to save money to purchase reusable nappies, rather than disposable ones.

The daily savings mothers would make by using reusable nappies could be better spent on other necessities. Mamma's Laef is also looking at ways it can lower the production costs of its products, thereby enabling it to lower its retail prices.

Scalability and future outlook

'We know we need to scale up [to meet the demand following the disposable diaper ban], but there is a considerable cost to do that.' – **BELINDA ROSELLI, TRUSTEE OF MAMMA'S LAEF**

Audio Visual education material: Mamma's Laef is investigating options to move its education and awareness material into video format. This will enable a broader range of communities to easily, and repeatedly, access the material, as well as minimise the cost of team members travelling domestically to provide the education.

Finances: Mamma's Laef is continually exploring long-term funding opportunities so it can continue training team members and distributing its products and education materials across Vanuatu. With additional funding, Mamma's Laef will also invest in industrial sewing machines and a larger working space to improve the efficiency of its manufacturing processes.

Two recent grants have allowed for a small number of care packages, containing reusable baby nappies, to be distributed to new mothers. These packages are seen as a valuable gateway for the ongoing use of reusables. Mamma's Laef is currently seeking funding to further expand the concept.

Ocean Sole

REDUCING OCEAN POLLUTION | REPURPOSING WASTE | SOCIAL ENTERPRISE

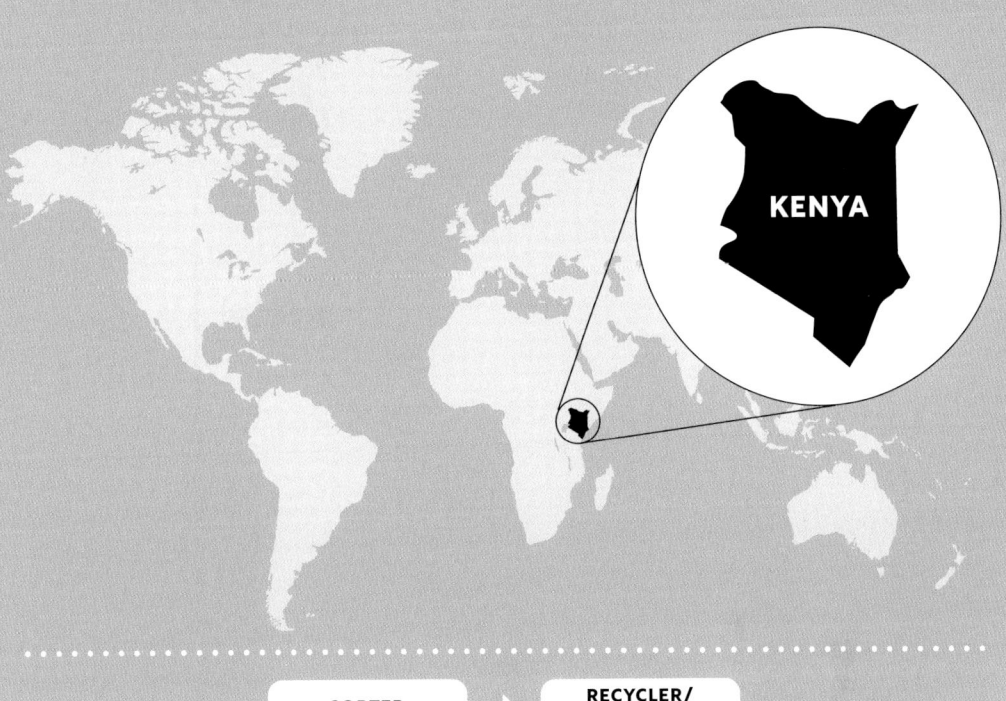

KENYA

SORTER → RECYCLER/ REPURPOSER

Flip-flops and beach litter, up to 1 million flip-flops per year, 1 tonne per week

Set up 8 years, operating since 2014

Multiple communities

CONTACT
W: https://oceansoleonline.com/ ● S: @OceanSole, #FlipTheFlop
E: solemates@oceansoleafrica.com ● T: +254 727 531301

ON THE ISLAND OF KIWAYUU OFF THE COAST OF KENYA, THE BEACHES WERE full of marine debris and plastic pollution. This pollution had accumulated to the point that returning sea turtles could no longer make nests to lay their eggs. The community on Kiwayuu Island decided to act. Women in the community started cleaning the beaches of litter, picking up toothbrushes, plastic bottles, fishing nets and flip-flops (thongs). Some mothers in the community took the discarded flip-flops home and created toys for their children.

It was watching these children play with their flip-flop toys on the beach one day that inspired Julie Church to start Ocean Sole, a social enterprise that turns flip-flop pollution into art to advocate for ocean conservation and to clean the beaches in Kenya. Ocean Sole was founded in 2006 and started with mothers selling their flip-flop art at local markets. By 2014, Ocean Sole had evolved into the business it is today, making thousands of flip-flop products each year.

Over 3.5 billion people, predominantly in the Global South, cannot afford shoes other than flip-flops. Due to the non-robust structure of flip-flops, people who wear them as their sole form of footwear often need to buy two to three pairs a year. This leads to billions of these non-decomposing plastic shoes being discarded into landfills, some of which are mismanaged, leading to the flip-flops 'leaking' into nearby beaches,

The Ocean Sole family with some of their colourful flip-flop creations. PHOTOGRAPH: OCEAN SOLE.

rivers, waterways, roadsides and the ocean. These environments are where Ocean Sole collectors recover the shoes and bring them to collection points to then be turned into unique artworks.

> 'Shortly after having a hands-on experience with picking up flip-flops, plastic and other forms of rubbish ... I began to realise the beach is but a reflection of the ocean.'

> 'Had I not gone to the beach clean-up hosted by Ocean Sole, I may not have gotten a crystal-clear understanding of ... why it is important to take part in community clean-ups and donate to environmental organisations.' – **ADZEAH WOLF, REFLECTION AFTER PARTICIPATING IN AN OCEAN SOLE BEACH CLEAN-UP**

How the program works

Collecting: Ocean Sole employs and engages a network of collectors from over 15 coastal communities along the Kenyan coast to clean beaches and waterways weekly. Collected flip-flops are delivered and sold by the kilogram to Ocean Sole. This payment supports many communities, individuals and their families who

Recovered flip-flops sorted into colour groups.
PHOTOGRAPH: OCEAN SOLE.

collect the waste. Collectors received approximately US$0.30 per kilogram of flip-flop pollution.

Cleaning: At an Ocean Sole workshop, the recovered flip-flops are cleaned, polished and sorted by colour by an all-women team.

Blocking: Ocean Sole artists select a mix of cleaned flip-flops and glue them together into a block, ready to be sculpted.

Carving: The flip-flop blocks are carved with knives and sanded into various artworks, particularly animal sculptures. Off-cut scraps from the block sculpting are collected and used to make mattresses for vulnerable girls and women in Kenya.

Finishing touches: The sculptures are given a final wash and finer details are then added, such as ears, tails, eyes and the signature Ocean Sole logo.

Quality checking and final product: The finished artworks are quality checked, tagged and packed ready to be sold. Profits from the sold artworks are used to enhance local Kenyan communities and continue the Ocean Sole product line and employment.

Resources

Finances: The production of Ocean Sole sculptures started small, with artworks sold at local markets. Over time, the sculpture process became more refined and the client market for the artworks grew. An initial large order of artworks from a client gave Ocean Sole the financial capital and market exposure to really get going and become the organisation it is today. Ocean Sole has not received any grants.

People: Ocean Sole has 150 full-time employees, including flip-flop sculpture artists and suppliers. The Ocean Sole management team consist of six people, including the chief executive officer (CEO). Most staff are based in Kenya, and some are based in Florida, USA, where the Ocean Sole USA fulfilment centre (i.e. packing warehouse) is based.

Community support: Ocean Sole is a social enterprise whose operations started at a community level, supporting and employing local community members from the beginning. As a social enterprise, Ocean Sole invests profits back into continuing its community programs that support and enhance family livelihoods and the conservation of local environments and wildlife. For example, the sale of a medium-sized sculpture supports one person in Kenya with basic needs, such as medical treatment, food and clean water, for a week.

Equipment: Not a lot is needed for Ocean Sole to operate. A workshop is needed with space to clean and block recovered flip-flops, as well as for artists to carve, sand and put the finishing touches on the artwork. To create sculptures, artists

An Ocean Sole toy being carefully shaped.
PHOTOGRAPH: OCEAN SOLE.

require knives for carving, sanders and glue to refine and add finishing touches. In addition to a workshop, Ocean Sole requires an office for accounts, sales, marketing and shipment administration activities. Fulfilment centres are located in the USA and UK.

> 'Every purchase of artwork supports the coastal communities to clean their beaches and provide finances to improve their livelihoods, such as affording food and education.' – **OCEAN SOLE STAFF MEMBER**

Environmental benefits

Since 2006, Ocean Sole has recovered and upcycled over 559 tonnes of flip-flop pollution in Kenya. On average, Ocean Sole recycles between 750 000 and 1 million flip-flops per year, with a record-breaking 65 000 artworks hand-made from these flip-flops in 2018. Ocean Sole collectors recover the flip-flops not only from beaches, but also from landfills, in-land waterways and roadsides. Basically, anywhere flip-flop pollution accumulates, the Ocean Sole collectors will find it.

In addition to collecting flip-flops, Ocean Sole directs a proportion of its profits into organising and funding community beach clean-up activities along the coast and nearby islands of Kenya. Every US$20 of revenue supports over 66 kg of plastic pollution recovered from the coastal environment.

Social benefits

Reducing unemployment: On average, Ocean Sole has grown by 50% per year. This growth has allowed Ocean Sole to increase its impact on reducing unemployment, cleaning beaches and advocacy by 80%. Over 1200 Kenyans are supported by Ocean Sole through employment and the collection of flip-flops. A minimum 15% of profits fund Ocean Sole community and marine conservation programs. From 2017 to 2019, sales revenue increased by 86%, which allowed Ocean Sole to open a new workshop and employ 25 more people.

'Looking directly into the eyes of your employees and knowing you can help them, and their families, is something we should all experience in life.' – **ERIN SMITH, CEO OF OCEAN SOLE**

Increasing welfare: Ocean Sole provides its employees with an employment welfare program. This program provides health care to workers and their families and assists with school fees, land purchases or emergencies by matching the financial savings workers put aside for those amenities. For example, over 200 children have received schooling from the wages and scholarships Ocean Sole provides.

Each year Ocean Sole also provides over 22 000 hot meals to its workers and conservation education programs to 28 schools in Kenya. In 2020, Ocean Sole partnered with the social enterprise Lapaire Glasses to provide corrected vision to 35 employees.

The off-cuts of toy production are used to make mattresses.
PHOTOGRAPH: OCEAN SOLE.

Women's welfare: Ocean Sole is striving to reduce its staff gender gap by integrating women across its entire value chain. Since Ocean Sole started, female employment has increased from less than 10% to now almost 50%. Ocean Sole operations also focus on supporting women's self-help groups around Kenya, such as the Kiwayuu Women Self Help Group, which collect and sell flip-flop pollution.

Over the past few years, Ocean Sole has started creating a new line of products from the flip-flop off-cuts created when each block is carved into a sculpture. These off-cuts are tightly packed together in a fabric casing to form a mattress for vulnerable women and girls in Kenya.

Barriers to success

Managing copy-cat organisations: Ocean Sole products are directly copied by other companies who maximise on Ocean Sole's brand by using the Ocean Sole story, images and impacts to sell copied products. There will always be market competitors, but copy-cat companies do not always follow the ethical social enterprise ethos that Ocean Sole is built upon. Ocean Sole encourages buyers to do their research to ensure that the flip-flop sculpture purchased is not supporting a business that is not Fair Trade.

Shipping costs: Shipping artworks from the workshops in Kenya to buyers in North America, Australia or the UK is challenging. Shipping is very expensive, and finding a financially viable shipment option is a barrier for Ocean Sole to expand its

Large dolphin sculptures produced by the Ocean Sole team. PHOTOGRAPH: OCEAN SOLE.

enterprise without decreasing the revenue returned to the communities it supports. To reduce regular shipping fees, Ocean Sole has opened fulfilment centres in the UK and USA. This has allowed Ocean Sole to also open an online store in the USA and fulfil orders from its in-country centre.

Minimising workshop and training costs: Setting up new workshops and training new staff come with their own costs. As the Ocean Sole enterprise expands, it is looking to perfect its enterprise model so these costs are minimised and revenue back into the communities is maximised.

'Giving back to the community is a firm foundation to our mission ... We have a strong team of both men and women who come from all walks of life and each person has a different skill, talent or ability that we help nurture and develop so that one day, they can do things for themselves that once seemed impossible.' – OCEAN SOLE STAFF

Scalability and future outlook

As the production of plastic continues to grow exponentially, Ocean Sole plans to combat the global plastic pollution crisis by scaling its program beyond Kenya and into the Global South. Ocean Sole wants to take its successful enterprise to other underprivileged coastal communities that are negatively affected by plastic pollution so that these communities can benefit from the environmental and social gains that the program offers.

Over the next year, Ocean Sole will perfect its flip-flop upcycling model in Kenya and then establish its first international operations, focusing on Haiti, Indonesia, Brazil, India and Egypt, as well as more broadly in Central America and the Caribbean.

PET Recycling Company NPC (PETCO)

PRODUCER RESPONSIBILITY ORGANISATION | VALUE CHAIN SUPPORT | COLLECTION AND RECYCLING

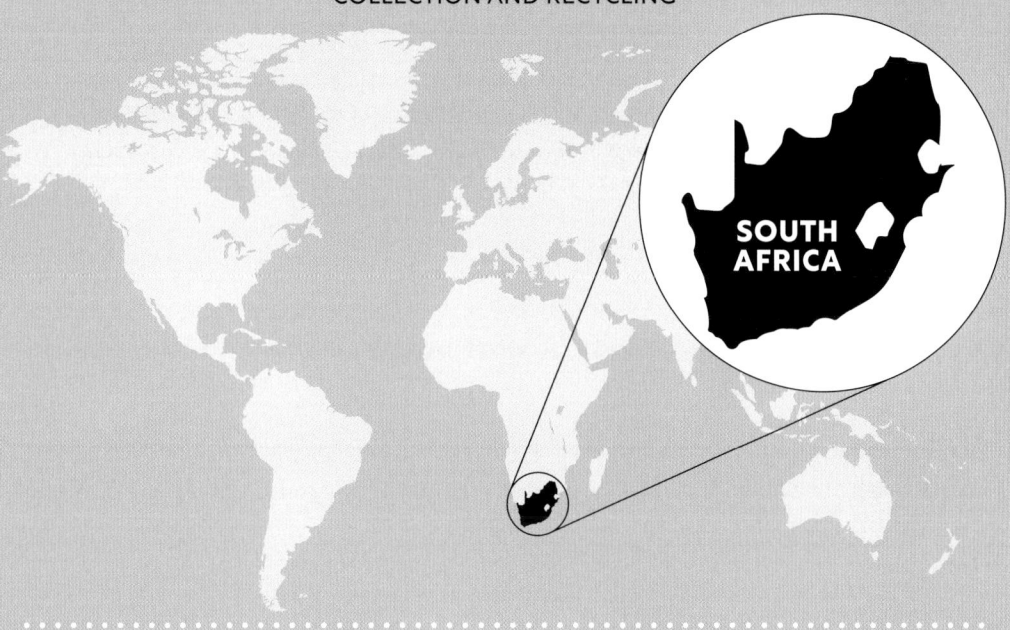

SOUTH AFRICA

BROKER

7533 tonnes PET per month

Set up 4 years, operating since 2004

Countrywide

CONTACT

W: https://petco.co.za/ • YouTube channel: https://www.youtube.com/user/1isPET
S: @PETCO_SA; @PETPlasticRecyclingSA • E: info@petco.co.za

IN THE EARLY 2000S, AS THE 'POLLUTER-PAYS PRINCIPLE' BEGAN TO INFORM waste management around the globe, South Africa introduced regulations relating to technical specifications for particular products, such as plastic bags. Embracing the principle, the polyethylene terephthalate (PET) industry within South Africa formed a working group across all sectors of the PET value chain and implemented a voluntary extended producer responsibility (EPR) model. The model would rely on PET producers paying a voluntary fee that would fund the collection of their PET products, predominantly plastic bottles, and fund the recycling of the PET into recycled PET (rPET) for plastic bottles and fibre.

The PET industry understood that for their EPR model to work they would require a viable end-use market to drive the collection and recycling of PET bottles. The working group identified and invited a company to set up a recycling plant that would be dedicated solely to post-consumer PET recycling. The plant became operational in October 2000.

'Plastic bottles have the incredible potential to create a circular economy.' – **PETCO TEAM MEMBER**

The establishment of the voluntary EPR model led to the working group formalising in 2004 as a not-for-profit organisation called the PET Recycling Company (PETCO), an industry-driven and financed national recycling initiative for PET plastic. PETCO is a vehicle through which the PET industry can coordinate its product stewardship and recycling activities. PETCO's mission is to minimise the environmental impact of post-consumer PET on the South African landscape. It does this by growing the collection and recycling of post-consumer PET, along with the labels and closures that accompany PET bottles, supporting existing and

PETCO sponsored clear bags to the Litter 4 Tokens NPC program. This is a program where recyclable materials are swapped for tokens which can be used to redeem household necessities. PHOTOGRAPH: PETCO.

PETCO sponsored the Timele Greening Project in Luphisi, Mpumalanga, with push trolleys to facilitate the collection of recyclables. PHOTOGRAPH PETCO.

encouraging new collection and recycling initiatives, as well as through public education and awareness campaigning.

By 2009, the first food-grade rPET was produced in South Africa and, 2 years later, with support from PETCO, a bottle-to-bottle PET recycling plant approved by Coca-Cola South Africa was established by PETCO's recycling partner, Extrupet Pty Ltd. In 2020, Extrupet completed the expansion of its food-grade plant, which increased the bottle-to-bottle recycling capacity to just under 3000 tonnes per month, or over 30 000 tonnes per annum.

The plant expansion strengthened South Africa's position as a leader of the circular economy in Africa and allowed for PETCO to be competitive with other stakeholders in the global packaging market.

How the program works

PETCO is a producer responsibility organisation (PRO) that acts on behalf of its producer members to fulfil their mandatory EPR obligations, as per South African law. PETCO also supports the PET recycling system in different ways, as described below. PETCO was initially established as a voluntary PRO, but has been operating under mandatory EPR since 2021.

Mandatory EPR fee: Initially, the EPR fee was established in 2015, with the support of seed funding provided by Coca-Cola South Africa. The voluntary fee was the key component to PETCO's system and was paid by PET industry companies that joined the PETCO system (then known as PETCO's Voting Members).

In the previous model, companies that purchased PET resin (virgin and recycled) or preforms/sheeting that was either produced locally or imported had to pay a per-tonne fee (currently R620 per tonne). In addition to the fee, some brand owners contributed an annual grant to support PETCO's programs. In line with best international practice, PETCO now levies EPR fees based on the amount of product that its producer members place on the market.

However, in 2020, Section 18 of the *National Environmental Management Waste Act* was introduced making it mandatory for obligated producers to: join an existing PRO and pay mandatory EPR fees; form a new PRO; or develop and submit an independent EPR scheme for their packaging (identified products) to the Department of Forest, Fisheries and Environment.

Recycling value chain financial support: To ensure that a consistent demand for post-consumer PET from collectors persists, PETCO uses its revenue to help PET recyclers, particularly during adverse economic cycles. To maximise the positive impact PETCO can make with its financial resources, it closely monitors global and local market prices for virgin PET and rPET and adjusts its level of support to recyclers on a quarterly basis. This way PETCO's financial support can maintain its effectiveness by assisting the recycling chain more when market prices are low and less when market prices are high.

Supporting the collection network: The collection of post-consumer PET is largely completed by a network of informal collectors and micro-entrepreneurs who collect PET bottles, along with their labels and closures, from comingled waste. These collectors deliver their recovered PET bottles to industrial-scale recycling operations, typically via an intermediary agent such as buy-back centres.

In many regions of South Africa, PET bottle collections provide an income opportunity and improve the livelihoods for the urban poor. PETCO supports these collectors by providing them with equipment to collect the bottles, a safe place to separate and store the bottles, safe working conditions and help with efficient baling and transportation of PET bottles to recyclers. In partnership with municipalities and other organisations, PETCO empowers informal collectors through training and mentoring in entrepreneurship and enterprise development.

Educating consumers: PETCO runs education and awareness campaigns for the public of South Africa. These campaigns are tailored to the current attitudes and behaviours of the community. For example, in 2020 PETCO ran an education campaign on waste collecting as a trustworthy, legitimate form of income, and in 2022 PETCO began a campaign to encourage consumers to separate their recyclables from waste.

'We want to support the informal sector, destigmatising waste collectors and the important work they do.' – **PETCO**

PETCO also runs education campaigns that encourage community recycling and the recognition of plastic bottles as a resource, not garbage. In 2019, research found that one-third of South Africans did not recycle due to a lack of space in their house and confusion about where to deposit the recyclable material.

Resources

PETCO is not involved in the physical collection or recycling of PET. However, it acts on behalf of its members and uses their resources to drive recycling activities, including supporting waste collectors, particularly from the informal sector, guiding product design and building local recycling infrastructure to keep the PET value chain continuing in an efficient and affordable manner.

Finances: The PETCO system relies on annual mandatory EPR fees paid by producer members.

People: PETCO is an organisation made up of people representing all sectors of the PET value chain. PETCO consists of a board of directors (each director representing a different sector of the value chain), members (all companies that pay the EPR fees set by PETCO) and other partners who subscribe to PETCO principles but do not place packaging onto the South African market and therefore do not pay the EPR fee. PETCO currently has 18 paid internal staff members and four contracted staff to run the national program.

Infrastructure: PETCO does not own any infrastructure, but does finance the provision of infrastructure required for the improvement of the recycling chain. PETCO finances the equipment for waste collectors in the informal sector to improve the quality, quantity and efficiency of PET collection. For example, in 2021 PETCO, in collaboration with partners, provided waste

PETCO in partnership with Polymer Producer Safripol sponsored a trailer to Isphepho Envio Ambassadors to assist with the collection of recyclables. PHOTOGRAPH: PETCO.

collectors and buy-back centres with 18 trailers, 74 trolleys, 19 scales, 424 bins and 3570 bulk bags to support their collections.

PETCO also provides collectors with personal protective equipment and baling machines. PETCO also supports the recycling chain by assisting with the construction of local recycling infrastructure, such as buy-back centres.

In addition to the above, due to the impact of COVID-19 and the regulations introduced to curb the spread of COVID, 111 COVID-relief grants, 18 transport subsidies, 380 blankets, 20 620 separation-at-source bags and 10 top-up support subsidies were also provided to assist the value chain in recovering from the impacts of COVID-19.

Environmental benefits

By mobilising and supporting the PET value chain and informal waste collection, PETCO has significantly reduced the amount of PET waste entering the environment and has saved over 5 million cubic metres of landfill space. In 2000, before the formation of PETCO, only around 2% of post-consumer PET bottles were collected for recycling from landfill each year. In 2021, PETCO facilitated the collection of 63% of post-consumer PET bottles placed on the market by PETCO members, equating to 90 402 tonnes of PET bottles collected. This was an incredible achievement in a relatively short time frame.

Barriers to success

Increasing recycling targets and end-user markets: Unlike plastics, the reuse of steel, cardboard, paper and glass used in packaging is helped enormously by the ability of the original manufacturers to recycle their own material recovered from the waste stream. Polymer manufacturers are unable to offer such a low-cost route because of the variations in grades of polymers produced. Companies that recycle polymers must compete with the variable market prices of virgin polymers.

For PETCO to continue to grow its system and increase the volume of PET collected and recycled, it needs to increase the recycling capacity of its current partners, particularly the bottle-to-bottle system, and consider partnering with other recyclers and investors. rPET capacity needs to increase by two- to threefold by 2025 for PETCO to meet the ambitious rPET product targets set by the government, PETCO members and other non-governmental organisations in South Africa.

The success of the growth of the PET recycling industry is dependent on growth in the end-user market for rPET. The market demand for rPET will inevitably dictate the pace at which PETCO and the PET recycling industry can increase their capacity to meet market demand.

Designing for recycling and a circular economy: Another challenge for PETCO that constrains the quantity and quality of PET recycled is producers making plastic bottles that cannot be recycled into food-grade rPET material or that cannot be recycled at all. When it comes to bottle-to-bottle recycling, this means PET that could be recycled and turned into new bottles is not recycled, and is therefore inevitably destined for landfill. For example, bottles that are brightly coloured or have shrink sleeve labels cannot be recycled into food-grade PET.

PETCO is working with manufacturers to assist with the design of their bottles so that they can be successfully recycled into new bottles, thereby creating a circular economy of bottle-to-bottle recycled products. In addition, PETCO is looking for alternative end-use markets for PET bottles that cannot be recycled into food-grade or bottle-to-bottle products. For example, heavily pigmented PET bottles can be used to make recycled plastic pallets that are used in the fruit and beverage industry, and green bottles can be made into strapping.

PETCO is stimulating the development of new local end-use markets as well as exporting rPET fibre that would otherwise end up in landfill.

Maintaining a competitive price: Globally, rPET prices are exceedingly increasing due to the demand for rPET being far in excess of supply. An increase in voluntary brand owner commitments as well as legislative instruments with explicit targets for the use of recycled content across the globe have compelled various sectors to reduce the use of virgin materials and increase the use of recycled materials.

Processed recycled PET plastic pellets. PHOTOGRAPH: PETCO.

The global shortage in the availability of rPET is mirrored in South Africa. In response to the need for increased availability in rPET, PECTO and partners are looking to invest in enterprise development initiatives that will seek to increase the availability of rPET. Coupled with this, attempts are also being made to improve the quantity and quality of PET bottles being collected to ensure that recyclers have adequate feedstock supply. These attempts include intensified support for the collection value chain as well as dedicated efforts to improve the design of PET packaging.

Adapting to changing legislation: A final challenge for PETCO is to adapt to changing legislation that affects its system. At the end of 2020, the South African Government introduced mandatory EPR for all 'identified products'. This legislation meant a change in the way PETCO operates in terms of structure, operations and membership.

Scalability and future outlook

PETCO contracts PET recyclers that operate in South Africa, but has also assisted with sharing the PETCO model in Kenya and Ethiopia.

The PETCO system shows how the EPR model can be used to maintain a recycling market for packaging material.

Because PETCO earns most of its funds from PET that is placed on the market by members and the demand for PET by consumers, the future of the organisation will depend on PET consumption. Managing potential drops in PET consumption and flow-on effects will require cooperation and collaboration from all actors along the PET recycling chain.

Plastic Bank

REVALUING PLASTIC | REDUCING OCEAN POLLUTION | GLOBAL REACH

EGYPT

PHILIPPINES

BRAZIL

INDONESIA

| COLLECTOR | → | SORTER | → | RECYCLER | → | RESELLER |

All plastic materials
except PVC,
Styrofoam and small
multimaterial layered
sachets

Set up 3 years,
operating since 2016

Global

CONTACT
W: https://plasticbank.com/   •   S: @PlasticBank
YouTube channel: https://www.youtube.com/c/plasticbank
T: Toll Free North America: +1 866 220 8474   •   T: International: +1 604 263 7443

PLASTIC BANK IS A FOR-PROFIT SOCIAL ENTERPRISE COFOUNDED BY DAVID Katz and Shaun Frankson. Plastic Bank builds ethical recycling systems in coastal communities and reprocesses the materials for reintroduction into the global manufacturing supply chain. This reintroduced recycled plastic material is known as Social Plastic®. Plastic Bank aims to empower the world to stop plastic pollution while improving the livelihoods of those who collect it.

In 2013, during a tech conference on 3D printing and exponential technologies, David witnessed how 10 cents-worth of plastic could be reshaped into a product worth US$100. It was during this demonstration that David realised that plastic intrinsically has value. He called Shaun the same day and put the following idea into motion: '[To] reveal the true value of plastic, so that others realise that plastic is too valuable to discard into the ocean.'

David and Shaun combined their expertise in social enterprises and running large tech companies and set out to develop a program that prevents plastic waste from reaching the ocean, empowers marginalised communities and dispels the myth that plastic pollution found on beaches can only be downcycled into products.

Within 2 months, David and Shaun had amassed enough international and social media attention to generate a movement of millions of followers to spread awareness of plastic pollution and their mission. It took 3 years from the initial idea until the first Plastic Bank program was set up, with subsequent expansion into Indonesia, the Philippines, Brazil and Egypt.

'The only way to stop ocean plastic is to reveal the value in plastic by transferring as much value as possible into the hands of the collection communities.' – **DAVID KATZ, COFOUNDER OF PLASTIC BANK**

How the program works

Plastic Bank is a closed-loop recycling system that integrates existing local collection branches, infrastructure and processing partners into its system.

In each country where Plastic Bank operates, there is a local country manager and support team, who empower communities to run their own Plastic Bank-certified plastic recycling network and operations. The program involves inviting independent businesses already part of the local recycling system to join the Plastic Bank program, thereby circumventing competition between recycling businesses that already exist in the community.

For pre-existing collection branches and collection members in the community to be incorporated into the Plastic Bank program and be eligible for the rewards

A member of the Plastic Bank team collecting plastic items on a beach.
PHOTOGRAPH: PLASTIC BANK.

bonus, they must adhere to Plastic Bank's code of conduct, not employ child labour, pass audits and inspections and agree to use Alchemy™, Plastic Bank's digital blockchain platform via the PlasticBank® app.

Collection community members are given a unique digital identifier, which is linked to the app. All transactions between members of the program are completed using the PlasticBank® app. Members can deposit plastic material recovered from the environment, or collected from households and businesses, at certified locations. The member's unique digital identifier is scanned via the app and the recovered plastic is sorted by material type and colour. Members receive the market price for the deposited plastic and an additional Plastic Bank rewards bonus.

Plastic Bank only provides payment to collection community members for plastic materials that their supply chain requires. However, if a local market exists for other recyclable materials, then members can also deposit those materials at the collection centres. For example, if the collection branch receives cardboard, organic and metal materials, then it can maintain the collection of those non-plastic materials.

The collected plastic waste is then sold to Plastic Bank-certified processors and recyclers. The plastic is recycled into Plastic Bank's Social Plastic® feedstock and then sold to Plastic Bank's branded partners, who then incorporate the Social Plastic® feedstock into their products. Plastic Bank's partners include Henkel, SC Johnson, Advansa, CARTONPACK and ScanCom.

The PlasticBank® app: The PlasticBank® app records and maintains all operations from a global standpoint. The app uses blockchain technology to allow each step of the Social Plastic® supply chain to be transparent, inspected and audited at any time to ensure all members are complying with Plastic Bank's code of conduct.

Via the app, Plastic Bank staff can track that all plastic is ethically collected and provide a fair and transparent payment system by validating the identity of all program members using a unique member ID. The app provides an auditable trail of every transaction and exchange in the system from the point at which the plastic is recovered from the environment through to the reintroduction of the plastic as a Social Plastic® product on a retail shelf.

Rewards bonus: The Plastic Bank program provides a bonus payment to collection community members on top of the market price they receive for the plastic waste they deposit at a collection branch. This bonus can supplement a member's income by up to 60% and helps them pay for basic necessities, such as groceries, cooking fuel, school tuition and health insurance.

Resources

Finances: The initial start up of Plastic Bank was financed by the funds David received following the divestiture of his global positioning system (GPS) tracking tech company. Since the initial financing, Plastic Bank has become a self-sustaining program. Revenue streams are built from selling Social Plastic® to branded partners, selling offset and impact programs to partners and conducting sponsored regional activities to expand infrastructure and meet client needs.

Community support: When Plastic Bank first introduced its programs, it hired an in-country team to appropriately set up and communicate the program to the local community. The in-country team also provided training to local community members so that locals could champion the running of the Plastic Bank program within their community.

Infrastructure: An aim of Plastic Bank is to use any existing recycling systems and infrastructure in the community and to work the program into these pre-existing systems, rather than competing or restructuring the systems. In countries with little pre-existing infrastructure in communities, Plastic Bank built the infrastructure and systems required for its recycling program to succeed and benefit the community.

Building new infrastructure is funded through an offset program, whereby branded partners invest in building the in-country recycling system infrastructure capacity for the community so that the community recycling system can meet Social Plastic® orders of the partners.

Plastic Bank uses an app to record the progress of all operations. PHOTOGRAPH: PLASTIC BANK.

People: Plastic Bank has its main headquarters in Vancouver, Canada, and an in-country team in each participating country. Plastic Bank employs over 200 people, including a mix of full-time staff and independent contractors, to run the program across four countries.

The key to Plastic Bank successfully establishing in a new country is its employment of an in-country team of staff who set up the program, train local community members and communicate the program. Using an in-country team is one factor that has helped Plastic Bank to establish over 500 collection branches in four different countries.

Training: Plastic Bank's Social Plastic® provides a direct, transparent supply chain from the recovery of the piece of plastic from the environment to that same plastic being recycled, shaped into a product and sold on a supermarket shelf. Plastic Bank provides training for partnering company staff so that the staff know how the Social Plastic® supply chain works and can convey that through their marketing and business. These trained staff can be proud that they work for a company that is doing something good.

Environmental benefits

Since Plastic Bank began, over 60 million kilograms of plastic pollution has been recovered from the environment.

The long-term goal is to reduce plastic pollution so that there is very little plastic to recover from the environment. The majority of collected plastic waste will be from community businesses and households that members can service, thereby creating a closed-loop recycling system in the community.

'Plastic Bank isn't about making people into life-long collection members, but rather optimises collection as a starting point for someone to get literacy training, access to technology and access to career training.' – **SHAUN FRANKSON, COFOUNDER AND CTO OF PLASTIC BANK**

Social benefits

By leveraging traditional and social media exposure, David and Shaun built a multimillion person following and took a proactive approach to position Plastic Bank as a solution not only for plastic pollution, but also for extreme poverty. The joint environmental and social benefits the Plastic Bank program provides to communities has enabled the program to successfully establish in a diverse range of countries and communities.

Plastic Bank provides a range of training programs for community members that not only help alleviate poverty, but also empower the community beyond finances. Additional programs include training in basic literacy, technology, health and safety, financial literacy and career advice.

Empowering women and communities: Traditionally, recycling is not a particularly female-driven industry. Plastic Bank intentionally targets women to register as collection members and operators, and employs them in the in-country staff teams.

The Plastic Bank program also empowers collection members to be acknowledged as part of the community. Plastic Bank's social programs provide collection members with their unique ID number, access to health and work insurance, fintech services and retirement funds.

Plastic Bank has recovered over 60 million kilograms of plastic pollution from the environment.
PHOTOGRAPH: PLASTIC BANK.

In addition to providing infrastructure that is required for the Plastic Bank program to operate, Plastic Bank helps communities by providing them with solar-powered Wi-Fi, lighting and mobile phone charging centres.

Barriers to success

Localisation: The biggest constraint in many countries rapidly adopting the Plastic Bank program is that the program requires localisation in each country. Fostering relationships and introducing the program into local cultures takes time. Plastic Bank invests a large amount of time and effort to ensure the program feels locally owned. Plastic Bank never wants to enter a community and set up a recycling or benefits program that cannot be supported or maintained.

Dispelling myths: During the 3 years it took Plastic Bank to set up its first program, David and Shaun wanted to dispel the myth that plastic pollution found along beaches could only be downcycled into products. They worked with partners to conduct pilot studies of beach-cast plastic found along the shorelines.

These studies showed that once recovered, cleaned and sorted into plastic material types, beach-cast plastic was of recyclable-grade quality. Plastic Bank also had to overcome the stigma that their program would not work because similar previous recycling programs had, in some part, failed.

Scalability and future outlook

Plastic Bank's goal, as it continues to expand into more countries, is to mobilise the program to grow exponentially. One way Plastic Bank is aiming to achieve this goal is by putting systems in place, introducing a licensing model, including partnership expansions, and optimising the app to provide digital training and onboarding. Reaching these goals will enable Plastic Bank to offer anyone in the world with the PlasticBank® app and a scale to start their own Plastic Bank program, recover and recycle plastic from the environment and earn rewards.

Global scalability: At the beginning, Plastic Bank's main objective was to ensure the program was scalable. If it could not be scaled, then the program would simply be a great marketing campaign and would not stop plastic pollution.

David and Shaun decided to first trial and establish the Plastic Bank program in disadvantaged regions. They figured if they could make the program successful, then scaling and transferring the benefits and success of the program to other regions of the world would be simpler.

Franchise the program: At its core, Plastic Bank was designed to be a globally exponential company. It was never designed to be a single-country program. Plastic Bank is now moving towards a license model, where any established plastic processor in the world can become part of the program. Processors can become a Plastic Bank-certified operation and benefit from connecting and working with Plastic Bank's partners and networks.

Transitioning to a license model will allow Plastic Bank to increase its plastic recycling capacity and the quantity of Social Plastic® it can supply to its partners, thereby reducing its partners' consumption of virgin plastic materials.

The Plastics Circle

WASTE REDUCTION | REPURPOSING WASTE | REVALUING PLASTIC

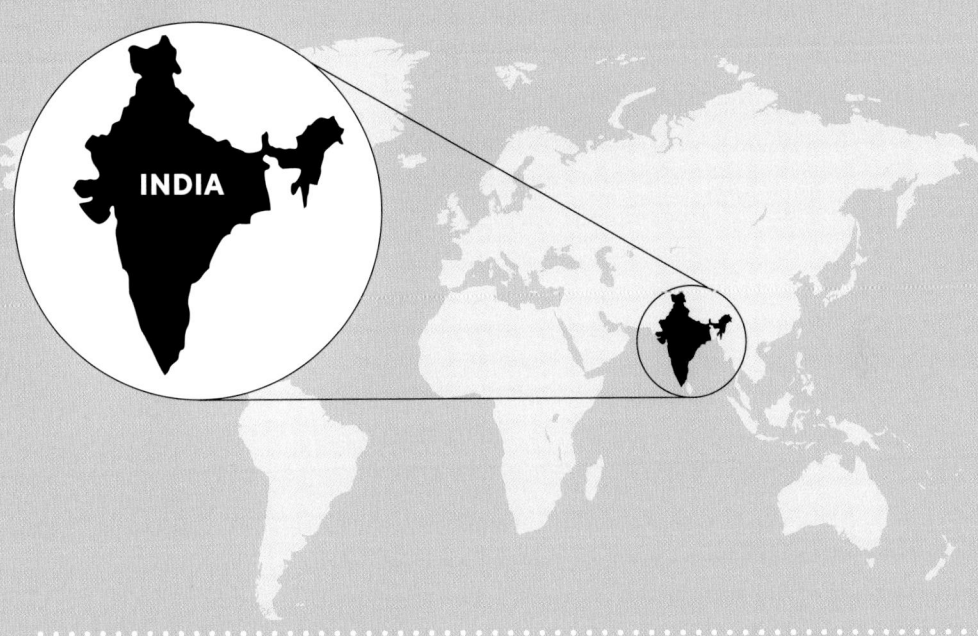

INDIA

BROKER

1200 tonnes of plastic
waste per year

Set up 2.5 years,
operation at pilot
stage

Multiple countries
within the Asia-
Pacific region

CONTACT

W: https://www.theplasticscircle.com/plastx • W: https://www.plastx.co
S: @PlasticsCircle • E: murray.hyde@theplasticscircle.com
L: https://au.linkedin.com/company/the-plastics-circle

WHAT HAPPENS WHEN YOU COMBINE A SCIENTIST, A BUSINESS MIND, A strategist and an operations expert and then apply them to solving plastic pollution? Answer: a new take on the problem that delivers an innovative solution.

Back in 2016, two of The Plastics Circle's four founders led a joint government and industry initiative, including creating a strategic plan for achieving packaging sustainability in Australia. Along the way, while researching global strategies, programs and initiatives, they discovered something critical was missing: the plastics circular economy. If society wanted to use plastic's benefits, the answer was a plastics circular economy, and yet there wasn't one!

A massive unmet demand for post-consumer-recycled (PCR) plastic was waiting for corporations to fulfil their commitments. However, in Asia, inefficient collection meant quality post-consumer-recycled plastic was scarce, and what was collected may have been tainted with modern slavery and material substitution.

The Plastics Circle fills this unmet market demand for quality post-consumer-recycled plastic in Asia, responsibly, with PlastX.

'How can we draw economically valuable plastics out of the system before they become waste and do this without traditional waste and recycling infrastructure?' – **TRISH HYDE, FOUNDER AND CEO, THE PLASTICS CIRCLE**

How PlastX Works

PlastX is an app where brands and plastic processors place orders for a specific material needed (by type, colour, condition, location and price). Registered collectors use the PlastX app to choose the jobs they want; they then take their collection to a hub (authorised PlastX retailer), where their identity is confirmed and the collection checked, logged, bagged and tagged. The data collected informs the route and frequency for logistics direct to the processor's door.

In Asia, the key to overcoming the unmet demand for quality post-consumer-recycled plastic is embracing the informal sector (not displacing them with infrastructure) to collect recyclable plastics to specification and provide them with a new level of income certainty.

To get the quality demanded, at market rates, The Plastics Circle's solution eliminates layers of middlemen and connects the collector to a hub and then to the processor. This is done with PlastX's proprietary physical and digital general ledger technology, which performs as a chain of custody, delivering quality post-consumer-recycled plastic economically and with verifiable provenance.

'How can we harness the potential of the informal sector in a responsible and transparent way to create economically viable plastic circularity?' – **MURRAY HYDE, FOUNDER AND CCO, THE PLASTICS CIRCLE**

Points of difference

Unlike other models for the collection of plastic to supply industry, PlastX is:

- demand- and quality-driven, which means plastic is recovered before it is contaminated and enters the environment
- an efficient three-step supply model that bypasses layers of middlemen, meaning more money in the hands of informal workers and a commercial model for the hubs, logistics, processors and The Plastics Circle
- a digitised people-based supply chain bringing together the people/partners needed to form the chain, and drawing traceable, responsibly sourced post-consumer-recycled plastic through to fill demand anywhere.

Pilot programs

PlastX recently piloted in Uttar Pradesh, northern India. To implement the program, The Plastics Circle worked with in-country partner Green Dream Foundation to operationalise the on-ground logistics of the PlastX program. Green Dream

An outline of how The Plastics Circle's PlastX program works. QA = Quality Assurance. Source: The Plastics Circle.

Foundation's local knowledge and language skills enabled them to locate appropriate sites for plastic collection hubs, recruit and train locally based collectors and collection hub personnel and to provide them with personal protective equipment and smart devices for the PlastX program. Green Dream Foundation also provided basic literacy and smart device education for collectors and hub personnel. The Plastics Circle's second pilot program in Thailand is just getting underway.

Resources

Finances: The Plastics Circle and PlastX program are for-profit (and purpose) commercial models. To date, PlastX has been funded through personal investment from each founder and, having been awarded a place in the Coca-Cola Amatil Xcelerate program, The Plastics Circle received an AU$38 000 investment to help establish its business. The pilot program in India was cofunded with Solidaridad Network, and the preparation for the pilot program in Thailand is being funded by The Incubation Network. The company is seeking seed funding for commercialisation and scale.

People: The Plastics Circle's headquarters are based in Sydney, Australia. The four founders work in the business as the chief executive, chief innovation officer, chief operating officer and chief commercial officer. The in-country operations for the pilot program in India were coordinated by part-time workers of Green Dream Foundation. In Thailand, The Plastics Circle has formed a partnership with a local organisation (GEPP) to act as its in-country partner for the upcoming pilot.

Community support: PlastX aims to provide collectors respect, fair treatment and income certainty that is free from abuse or exploitation, providing existing

Collectors recording information about the recycled products that have gathered. PHOTOGRAPH: THE PLASTICS CIRCLE.

businesses additional income as hub operators, and to recover plastic for a more efficient and sustainable plastic recycling system and cleaner community environment.

Using the local knowledge of in-country teams, the PlastX program and associated benefits were communicated to locals to determine the level of interest from pre-existing collectors, businesses and general community members who wanted to participate in the program. When the recruitment process for the pilot program was announced, all positions were filled (in fact oversubscribed) within two days.

Equipment: PlastX is a low capital expenditure model, which means it is inexpensive to establish, set up and run. It requires robust, well-developed software, secure data management and some office space for the digital platform that connects collectors, hubs and processors via basic smart devices and tablets.

Environmental benefits

PlastX's pilot program in India started with 10 collectors recovering almost 4000 kg of plastic in 67 collection days – almost 6 kg per person per day.

In Thailand, where global waste management giant, Veolia, is our foundational customer, PlastX will engage approximately 2000 collectors to remove up to 100 tonnes of plastic from the environment every month.

As additional customers come on board and PlastX scales, the attendant impacts for collectors and the environment grow exponentially.

A team meeting to upgrade collection information recording to digital devices. PHOTOGRAPH: THE PLASTICS CIRCLE.

Social benefits

Current post-consumer-recycled plastic supply chain systems feature up to nine layers of 'middlemen'. PlastX's digital platform streamlines this, directly connecting buyers with collectors via a hub, ensuring no other party can 'take a cut' of the revenue a collector receives from the material they collect. This direct connection allows the PlastX program to pay collectors more than they currently receive. For example, PlastX's collectors in India received 47% above the usual waste rate.

Partnering with Green Dream Foundation has ensured the PlastX program benefits the community by implementing a new waste collection system that improves on working conditions, hygiene and access to educational programs, as well as reducing child and forced labour, compared with previous systems.

Barriers to success

Engaging investors: Making change is complex, as is understanding and bridging the gap between the needs of informal waste collectors and those of brand and manufacturing partners. The Plastics Circle has found attracting investment from external parties for PlastX a challenge when the social and environmental issues on the ground in Asia are not well understood. The Plastics Circle is exploring different avenues for engaging investors and educating them on the trials faced by communities across the Asia Pacific and how the PlastX program is helping.

Regulation: Some countries in the Asia-Pacific region have laws preventing recycled plastic becoming food-grade plastic. This legislation restricts a significant market sector that The Plastics Circle can engage as customers of recycled plastic material.

Pre-existing systems: In parts of Asia, the network of scrap traders and brokers has a monopoly and exerts some control over the existing waste management supply chain. These pre-existing systems can discourage alternative supply chain systems from entering the market.

Scalability and future outlook

The Plastics Circle believes now is the right time for PlastX: major brands have commitments to use more recovered plastic and an inability to deliver. This, alongside European consumer expectations for brands to label post-consumer-recycled plastic content on packaging and concerns about the regulation of these claims, gives rise to an enormous potential market for PlastX, with its demand-driven, verifiable, efficient RecoveryTech platform.

As a digitised people-based model, PlastX is not limited by geography, making it a viable solution for any market where manufacturing demand exists in Asia.

With a contracted customer in Thailand, The Plastics Circle is preparing to establish a pilot program ahead of fundraising for commercialisation. Beyond Thailand, The Plastics Circle will look to other Asian markets, such as Indonesia, Malaysia, Vietnam and the Philippines, and a return to India.

Plastic Collective

COMMUNITY-BASED | REPURPOSING WASTE | GLOBAL REACH

AUSTRALIA

| COLLECTOR | → | SORTER | → | RECYCLER |

200 tonnes of
thermoplastic waste

Set up 3 years,
operating since 2018

Multiple
communities

CONTACT
W: https://www.plasticcollective.co/ ● S: @plastic_collective
L: https://www.linkedin.com/company/theplasticcollective/

THE BEGINNING OF PLASTIC COLLECTIVE IS A STORY OF THREE PIVOTAL moments in Louise Hardman's life that led her to found Plastic Collective. The first moment was when Louise was working as a zoologist and rescued a young green sea turtle stranded on the banks of a river. Unfortunately, the turtle did not survive and a necropsy (animal autopsy) of the turtle revealed that it had died with its entire digestive system full of plastic.

Many years later, Louise and her daughter were travelling through Thailand. They started a conversation with a 5-year-old girl after watching the girl pick up a bag of household rubbish and throw it into the river. Perplexed by what she had just witnessed, Louise queried the girl why she had thrown the bag. *'Everybody does it,'* she replied. *'It goes away.'* Louise understood that no one had ever explained to the girl or her community where 'away' was and the harm the rubbish could cause there. This experience spurred Louise to leave her zoology job and become an environmental educator in 2013.

It wasn't until 2016, when Louise was taking 3 months injury leave, that she felt that something in her life and work was not feeling right. Louise felt despondent, disempowered and unsure of what she should be doing with her life, and a friend suggested she needed to work with water. It was at that moment that the memories of the turtle, the plastics and the young girl in Thailand came flooding back to Louise and she understood what her next adventure would be.

Over the next 3 months Louise made a plan to empower herself and return to working with the ocean. She was to start a social enterprise that empowered and educated communities to turn waste into a resource, which would stop plastic from entering the ocean and would thereby protect the marine environment from the harms of plastic pollution. Thus, Plastic Collective was born.

Founder Louise Hardman with the filament produced by the Plastic Collective Shruder™. PHOTOGRAPH: PLASTIC COLLECTIVE.

'Many people feel disempowered thinking about the enormity of environmental issues like plastic pollution. I wanted to feel empowered, I wanted to empower myself with the solutions.' – **LOUISE HARDMAN, PLASTIC COLLECTIVE FOUNDER**

Two years later, Plastic Collective had prototyped, developed, implemented and refined the Shruder™ (pronounced shroo-der), a mobile machine that shreds thermoplastic waste and extrudes the shredded plastic into sellable recycled material that can be incorporated into new products. The Shruder™ is an adaptation of Dave Hakken's Precious Plastic Modular machine.

How the program works

Plastic Collective works with communities and their networks to build capacity for the recovery of recyclable material from kerbside bins and the environment. The material is sorted, cleaned, recycled via a Shruder™ and sold on to buyers and supply chains established by Plastic Collective in Malaysia, Hong Kong, the USA, Australia and Indonesia.

The first Shruder™ machine was piloted in Vanuatu in 2017, with its success prompting another four pilot studies in 2018–19 in Bali, Indonesia, Mantanani Island, Borneo and Australia. The pilot programs had varying levels of success. Plastic Collective took the lessons learned from the pilots and, in 2020, launched a more advanced hardware, software and training model called the Plastic Collective System, which included the Shruder™ Recycling Station.

The Resource Recovery and Recycling station is an enclosed shipping container that houses a range of equipment required to run a plastic recycling facility, such as a granulator, extractor, baler, generator, extruder and compression moulder. The station can be adapted to provide a community with variations of the Shruder™ Recycling Station, depending on their needs, and specialised training programs to establish enterprises and work with plastic materials. The first recycling station programs were implemented in communities in Bowraville and the Gulf of Carpentaria, Australia, and Bali, Indonesia in 2021.

Resources

Finances: The development and implementation of the Plastic Collective recycling program and stations were funded by an Australian Federal Government Cooperative Research Centres Grant of AU$2.4 million. The program was a partnership between Southern Cross University for monitoring, research and development and SouthPole for auditing, codeveloping and implementing the global plastic offset credit scheme '3R Initiative: Reduce Recover Recycle'.

Similar to carbon offset schemes, the plastic credit scheme allows businesses and individuals to offset their plastic waste production by purchasing plastic credits. The money from the plastic credits funds the provisioning and continuation of the Plastic Collective System programs.

Plastic Collective is working hard to include a sliding scale of entry-level costs to ensure the scheme is not financially prohibitive for small, remote-community recycling operations to participate. Plastic Collective does not want these communities excluded from the opportunities and benefits the scheme can bring to small communities.

Some programs are funded privately; for example, Gowings Bros Limited is funding the recycling program in Bowraville, Australia. Plastic materials from Gowings Bros shopping centres are collected and recycled through the Centre for Sustainable Solutions, with Gowings Surf businesses buying the recycled plastic to make products such as sunglass frames and surf wax combs.

People: Plastic Collective's core team is made up of five full-time and five part-time staff who run marketing, sales, media, accounting and administration.

Community support: The aim of Plastic Collective is to empower communities with knowledge and enable them with solutions. Plastic Collective focuses on communities that have poor infrastructure and high transport costs due to distances to large urban centres. Often, these communities import large proportions of their food and beverages, which arrive packaged in plastic, and do not have the financial capital to export the plastic waste. Empty plastic packaging is seen as valueless waste and is often burned, buried or dumped. Through Plastic Collective's program:

> '[t]hese communities don't see plastic as waste anymore. They see it as a resource...it has value, and that changes everything.' – **LOUISE HARDMAN**

Plastic Collective receives community support for their program because the program involves and is operated by the community from the beginning. For example, the Bowraville program is operated by the Women and Family Support organisation of MiiMi Aboriginal Corporation, who are setting up a Centre for Sustainable Solutions at the old Bowraville Butter Factory, transforming it into a recycling and skills community training centre.

The Gulf of Carpentaria program is operated by the Carpentaria Land Council Aboriginal Corporation (CLCAC) in partnership with Earthwatch Australia and their 'Wetlands not Wastelands' initiative, sponsored by the Coca-Cola Australia Foundation. The initiative locates leakage points of waste entering the mangrove and wetland environment and establishes collection systems and supply chains to recover waste from the environment and from kerbside recycling.

Infrastructure: For the program to operate, it requires a shelter with a solid, level floor; three-phase power to run machinery; a selection of equipment, including a baler, granulator and an extrusion mould; digital scales; and software applications. These are just the tools required for operation, but the key components to the program working are setting up the collection system and gaining knowledge of the supply chain, with a focus on plastic pollution and plastic in kerbside recycling.

Each resource recovery program operates using specific Plastic Collective software technology to provide a transparent, auditable supply chain with provenance tracking. Every batch of plastic recycled by a community has a unique quick response (QR) code that can be scanned at each transaction point along the supply chain. The software allows Plastic Collective, the community and brands to track the progress of the recycled material.

Ethical and environmental compliance: All Plastic Collective programs are required to comply with Plastic Collective's ethical and environmental code of conduct to ensure no operation abuses workers, involves child labour or creates harm in the environment. For example, recycling operations cannot have leakage of microplastics into the environment and any recovered non-recyclable material must be disposed of properly in an appropriately managed landfill.

Training course: The Plastic Collective System includes a comprehensive training and education program, with eight modules. The modules are specific for different roles and positions, such as management, collection crew, educators and machinery operators.

Environmental benefits

The Resource Recovery Stations aim to significantly reduce the amount of plastic waste within communities. Each recycling station has a target to recycle 200 tonnes of material per year, with a particular focus on plastics with high recycle value, such as polyethylene terephthalate (PET), high-density polyethylene (HDPE) and polypropylene (PP), as well as lower-valued soft plastics.

The reductions in plastic waste in communities have also improved the aesthetics of these areas and made them more desirable to visitors. For example, following the pilot program on Les Village, Bali, the mayor said that the village had

Founder Louise Hardman with one of the Plastic Collective Shruders™.
PHOTOGRAPH: PLASTIC COLLECTIVE.

a new fragrance now that they had stopped burning plastic as their main form of waste management.

Social benefits

Elimination of shame: The biggest social benefit Plastic Collective has noticed from its programs is that workers in the resource recovery and recycling programs no longer feel ashamed about essentially managing rubbish. When Plastic Collective explained the value of the discarded materials to the community and the opportunities that extend from harnessing that value, workers and communities became proud of their recycling program.

Local employment: The Shruder™ recycling station program is operated by the local community, with the machinery operations of each station employing between 5 and 12 people. Community members are trained to operate the equipment and business management staff are educated on plastic pollution and its impacts.

Generating new opportunities: The revenue from selling recovered material has enabled communities with the financial capital and skill set to run other community employment opportunities and businesses. When Plastic Collective first approaches a community, it asks about any other social and environmental activities the community is undertaking or wants to undertake. Plastic Collective takes these additional activities into consideration when tailoring the recycling program to the community to ensure more widespread benefits.

Barriers to success

The Shruder™: The original Shruder™ design had low capacity (10 kg plastic per hour), partly in an effort to keep the machine running on single-phase power. Because of the low capacity, it took communities long hours to shred and extrude plastic, meaning the system was not economically viable. To overcome this, a second version of the Shruder™ recycling station was developed that runs on three-phase power and has a much higher capacity (≥150 kg plastic per hour).

Hot weather: For operators in tropical areas, working in the Shruder™ Recycling Station shipping container is very hot and humid, particularly in the tropics of the Gulf of Carpentaria. To make the working conditions in recycling station tolerable in these hot climates, Plastic Collective has adapted its container so that it can be housed under a roofed open space to allow maximum ventilation while continuing to protect the machinery against the weather.

Training: During one pilot study, the team receiving the Shruder™ did not want to partake in the training. This led to the team not being able to extrude plastic correctly from the equipment and needing to stop operations. Although the training takes time, Plastic Collective encourages all recipients to undertake the training so that they can maximise the efficiency of the Shruder™.

Source separation: Separating recyclable materials at the source is key to enable better efficiencies for the whole recycling system. In Australia, most recycling is placed into one comingled recycling bin. This comingling of material leads to high contamination rates and therefore low material recycling rates. Implementing a separation-at-source process would greatly improve the volume, quality and types of materials that can be re-entered into the manufacturing supply chain.

Government grants: Initially, obtaining grants was difficult because most government grants available for business start-ups required the grantee to match the funding amount granted to them. In time, Louise was able to match small amounts of funding with her own personal savings and money she borrowed from friends and family.

Getting government support: In Australia, Plastic Collective had interest from many local governments to introduce the Shruder™ Recycling Station. However, many were apprehensive about breaking contracts with current recycling and waste management companies. This has limited how quickly Plastic Collective has expanded its program across Australia, although local governments that manage their own waste collections are now seeking proposals.

'Local governments need to change the mindset from waste management to recovering resources.' – **LOUISE HARDMAN**

Scalability and future outlook

Plastic Collective system programs are currently operating in Association of South-east Asian Nations (ASEAN) countries. Plastic Collective is aiming to expand operations into more regions across multiple countries, particularly countries where plastic leakage and accumulation are highest. Plastic Collective predicts interest and participation in the plastic credit scheme will grow, and this growth will fund the expansion of recycling programs in new regions.

A recycling station set up at Miimi Aboriginal Corporation at Bowraville. PHOTOGRAPH: PLASTIC COLLECTIVE.

Plastic Free July®

EDUCATION | CAMPAIGN | GLOBAL REACH

AUSTRALIA

REPLACER

300 million kilograms
of plastic pollution
each year

Since 2011

Global

CONTACT
W: https://www.plasticfreejuly.org/   •   S: @plasticfreejuly
L: https://www.linkedin.com/company/plastic-free-july

IN JUNE 2011, AT A RECYCLING FACILITY IN PERTH, REBECCA PRINCE-RUIZ SAW a mountain of waste that her suburb in Australia had thrown away and witnessed the complex operations involved in processing it. Sure, recycling is important but, in that moment, Rebecca knew the best thing she could do to make a difference was to reduce her waste. The enormity of it shocked her into committing to a personal challenge that has grown into a global movement to end plastic waste. Rebecca challenged herself to try to avoid single-use plastic for the following month.

At the time, Rebecca was working in waste and sustainability education in a local government in Perth, Australia. Others from her community soon decided to join in the challenge. Each week, the community group sent emails that shared ideas and discussed solutions and alternatives. In 2011, people weren't talking about single-use plastics and waste avoidance, but by doing this challenge together they were able to share ideas and support each other to make change.

The first challenge started with a small group of 40 people (Rebecca's family, friends and colleagues) in the city of Perth. Following this, the challenge grew. People shared their experiences with others, as they felt good about reducing their waste and making a difference. In 2012, Rebecca set up a social media account, and the following year a website, to share ideas and actions.

From the start, people taking part in the challenge always focused on solutions, rather than the plastic pollution problem, such as purchasing and using reusable bags, water bottles and coffee cups, refusing plastic straws, choosing to buy unpackaged fresh produce, switching from liquid to bar soap or choosing a bamboo toothbrush or beeswax wrap over the plastic alternative. The group shared solutions on how to prepare food without packaging, where to purchase alternatives and ways to remember to do things differently. After the first year, people took Plastic Free July beyond their homes and into their schools, community groups and workplaces.

The activities of Plastic Free July became as diverse as the participants: from produce free of plastic packaging at farmers markets in Fremantle, Australia, to Vancouver Island in Canada going plastic bag free; to schools in New Zealand and England making beeswax wraps; to community groups in India making reusable fabric shopping bags. Around the world, customers tagged their local cafes in a social media post to inspire the cafes to set up 'mug libraries', a program that provides clean second-hand coffee cups to customers instead of single-use cups. Elsewhere, libraries organised practical cooking workshops, displays and competitions that promoted 'waste free' behaviour. The state of New York proclaimed the month of July as 'Plastic

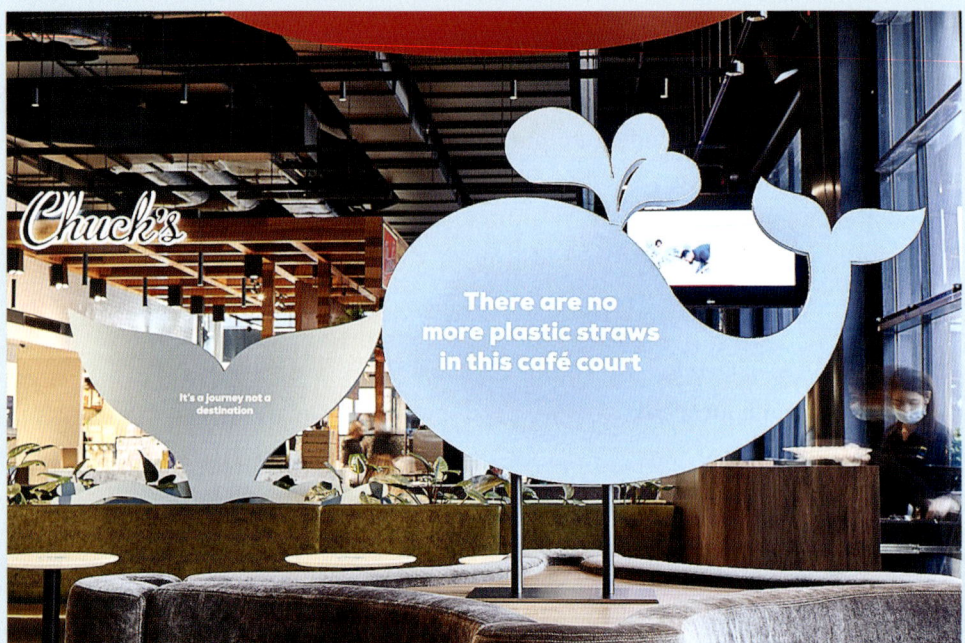

An example of organisations making lasting changes through their participation in the Plastic Free July challenge. PHOTOGRAPH: BROOKFIELD PROPERTIES.

Free July', with corporate organisations also participating through staff engagement initiatives, auditing the plastic used in their operations, making changes to procurement practices and supply chains, and big brands changing their packaging.

Since the first challenge back in 2011, Plastic Free July has become a global movement with a 140-million-strong community across 190 countries, empowering people to reduce single-use plastic consumption, one choice at a time, to create a cleaner future.

How the program works

Plastic Free July is a global behaviour change campaign that helps millions of people be part of the solution to plastic pollution so we can have cleaner streets, oceans and beautiful communities. Over the past decade this individual behaviour change has been the seed of cultural and systems change.

Each year, for the month of July, people and organisations can sign up to participate in the Plastic Free July challenge to find alternatives for the single-use plastics they use every day. Plastic Free July provides people with resources and ideas to reduce single-use plastic waste every day in their homes, workplaces, schools and communities.

Due to its global reach and incredible uptake in participation, Plastic Free July is now delivered by Plastic Free Foundation, which is an independent not-for-profit group. The role of the Plastic Free Foundation is to fully support the global Plastic Free July campaign and deliver its purpose of helping end plastic waste. Through advocacy and initiatives like the Plastic Free July challenge, the Plastic Free Foundation stops approximately 300 million kilograms of plastic polluting the world each year.

Resources

Finances: The initiative did not require any funding to get started. Initially supported through local government, the campaign team grew as part of Rebecca's part-time role in local government and through volunteer efforts. The campaign rapidly grew and scaled-up, building the tools and resources to a point that, in the second year, groups around the world were implementing Plastic Free July in their own communities.

Now operating through the Plastic Free Foundation, the program has been funded through a combination of government grants, corporate and philanthropic partnerships, memberships and donations.

People: The Plastic Free Foundation is a small organisation with four part-time staff and several supporters and volunteers.

Community support: From the very first day, Plastic Free July received enormous community support after being seen as a positive initiative offering something for everyone. Every year, millions of people participate and share the challenge in their communities. In 2021, people from almost every country in the world signed up to Plastic Free July.

Replacing disposable bowls with reusable bowls at a festival. PHOTOGRAPH: PLASTIC FREE FOUNDATION.

Over the past decade, awareness and concern about the plastic pollution problem has grown exponentially. At the same time, people's behaviours have not always matched their concern, and the challenge provides people with choices to do something about this. Plastic Free July offers practical solutions and ideas for people from around the world to take action, and shares stories of what other people and communities are doing. For 5 years the Plastic Free Foundation has worked closely with a behavioural economist to make sure its messaging is positive and that the campaign is inspiring and empowering to the general public.

The Plastic Free Foundation also plays an important role in surveying public attitudes to plastic and reporting the results to government to engage with the development and implementation of regulations, including single-use plastic bans and container deposit schemes. In late 2021, the Plastic Free Foundation partnered with Ipsos and WWF to conduct the first comprehensive global polling on actions to combat plastic pollution. The global survey revealed that nearly 9 in 10 people think it's important to have a global treaty to combat plastic pollution, which subsequently was agreed upon at the UN Environment Assembly in 2022. Continuing to share the general public's expectations for action to end plastic pollution will be important in moving forward future initiatives.

Environmental benefits

Although Plastic Free July is focused on single-use plastics, the Plastic Free Foundation's research has shown that when people take action to reduce their plastic waste, other waste such as food waste is also reduced. Globally, in 2021, participants reduced non-recoverable (landfill) waste by 1.2 billion kilograms and recyclable waste by 900 million kilograms, including a reduction in plastic consumption by 300 million kilograms. After 11 years, Plastic Free July has reduced the global market of bottled water by 2.3%, fruit and vegetable packaging by 3.1% and plastic straws by 4%.

By leveraging Plastic Free July, many companies, both big and small, have started to look for ways to eliminate and reduce single-use plastics in their operations and supply chains. Some of the major brands and companies announcing changes to reduce plastic waste during Plastic Free July include the Heineken Co. in Europe, Chubb and Fuji Film in South-East Asia, Colgate-Palmolive in the USA and Air New Zealand.

At an individual level, on average almost 90% of participants who completed the challenge made at least one lasting behavioural change, such as choosing to refuse

Reusable glasses being used at a market to reduce plastic waste. PHOTOGRAPH: PLASTIC FREE FOUNDATION.

single-use plastic bags. Plastic Free July participants reduce their household waste and recycling by 15 kg per person per year.

Social benefits

Increased social well-being: The Plastic Free Foundation measures the annual impacts of Plastic Free July through surveys of participants around the world on their waste attitudes, behaviours and creation. They also survey a general population in Australia. The survey results, published in reports available on the Plastic Free Foundation website, have shown challenge participants feel more socially connected and have a higher sense of well-being than non-participants.

Sense of achievement: Given the concern about plastic pollution in the general community, Plastic Free July offers people everywhere the opportunity to take action and be part of the solution. Plastic Free July appeals to every demographic, is apolitical and has something for everyone: it's about doing what you can where you are. For individuals and organisations, there is great appeal to being part of a global movement that is bigger than yourself, and many organisations, including non-governmental organisations, businesses, community groups and regulators leverage off this.

Barriers to success

Capacity building: The main constraint on the program is a lack of capacity to fully support participants and scale the behaviour change campaign to a broader global audience. Securing funding and building the capacity of the small Australian-based Plastic Free Foundation team is a challenge. With a new bold strategy to 'Help end plastic waste' and clear objectives, the Plastic Free Foundation team is working to build a sustainable entity in supporting global participants to deliver behaviour change in different countries. Waste avoidance is key to tackling the plastic waste issue, but often resourcing focuses on clean-ups and recycling. Turning off the tap to plastic waste is critical, but is a large resource challenge.

Scalability and future outlook

Through Plastic Free July, the Plastic Free Foundation has a vision of a world without plastic waste. After a review of their proposition, they have a new 10-year strategy with clear ambitions and are looking for aligned partners to scale the Foundation globally. Some of the ambitions include:

- 100 million consumers to sign up for Plastic Free July
- most consumers refusing single-use plastics
- big retail and consumer brands adopting plans to be plastic free
- governments globally legislating to ban plastic waste.

Plastic Free July is eminently scalable, with incredible opportunity to create new social norms around waste avoidance in different contexts, cultures and countries. Given the predicted increases in global plastic production, key parts of the Foundation's global strategy include waste avoidance, reducing plastic production at the source and changes in the design, use and reuse of plastics to create a circular economy.

Humans currently use as much ecological resources as if we lived on 1.75 Earths. Plastic Free July can be a powerful starting point to invite consumers to be less wasteful and to consider their consumption for a cleaner future.

PRISM Bangladesh Foundation

MEDICAL WASTE MANAGEMENT | SANITATION | HEALTH IMPROVEMENTS

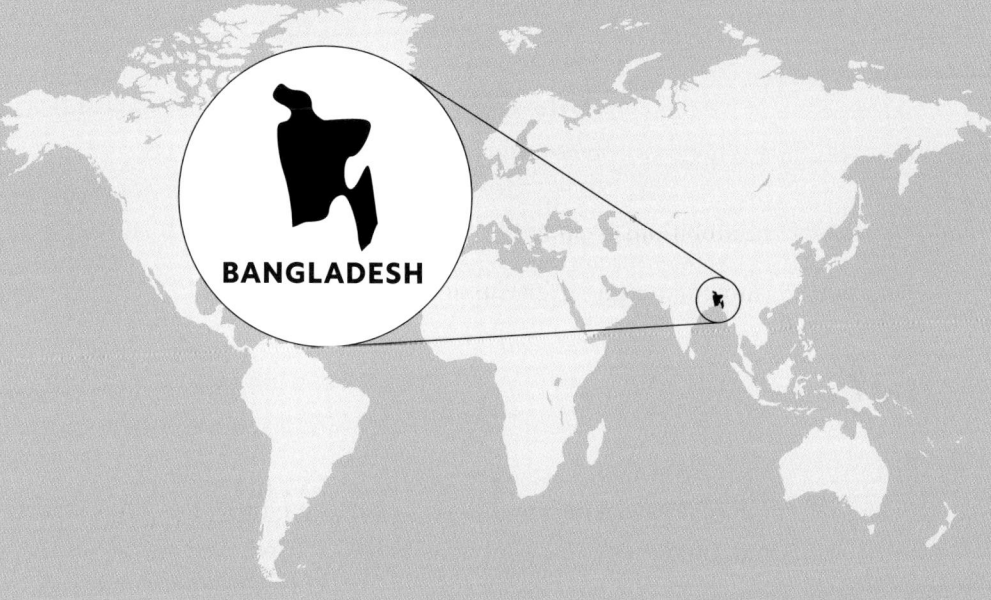

BANGLADESH

COLLECTOR → RECYCLER

21 000 kg medical waste per day, including infectious material, pharmaceuticals, sharps, recyclables and body tissue

Set up 1 year, operating since 2005

Multiple regions

CONTACT

W: http://pbf.org.bd/programs/Medical-Waste-Management ● S: @prismbdfoundation
E: prismbdf@yahoo.com; info@pbf.org.bd

THE PRISM BANGLADESH FOUNDATION (PRISM) IS A NON-FOR-PROFIT organisation that was established in 1989. PRISM's aims are to reduce poverty and enhance the socioeconomic conditions of rural communities within Bangladesh, particularly those with landless, marginalised individuals. Over the years, PRISM has implemented a wide range of development programs, including water supply and sanitation, rural enterprise formation, disaster relief and rehabilitation, as well as sustainable environmental development and management.

One environmental problem PRISM is solving is the sustainable disposal of medical waste. Until 2004, no authorised proper medical waste collection, treatment plant or dumping facility existed in Bangladesh. In major urban areas of Bangladesh, healthcare establishments such as hospitals would discard their infectious, recyclable, sharps and pharmaceutical waste in general household waste roadside bins.

The presence of such hazardous waste along city roadsides was very dangerous for community members, particularly waste collectors, and polluted the nearby air, water and soil. Many community members experienced health problems due to the waste.

It was a dream of Khondkar Anisur Rahman, Executive Director of PRISM, to improve the disposal of medical waste and thereby improve the health of the Bangladesh community. In 2005, PRISM piloted its first medical waste management program in Dhanmondi, a prominent residential area in the capital city Dhaka, where many private hospitals and clinics operate.

A truck used to transport medical waste in Rajshahi, Bangladesh. PHOTOGRAPH: PRISM.

PRISM's pilot program initially collected, treated, disposed or recycled medical waste from 17 healthcare establishments in Dhaka City. The program has expanded and now services 1121 healthcare establishments across six districts, including Narayanganj, Rajshahi, Rangpur, Sylhet, Jessore and Savar.

How the program works

The PRISM medical waste management program is a safe collection and final management system for hazardous medical waste. Healthcare establishments pay a service charge for PRISM to collect and properly manage medical waste. The healthcare establishments segregate medical waste into three streams using separate bins: infectious waste, plastic waste and sharps waste. PRISM collects all three waste streams and delivers them to its waste treatment facility.

Plastic waste is chemically disinfected, shredded and sold on to scrap recyclers. Infectious waste is sterilised using an autoclave and then incinerated. Sharps waste is sterilised using an autoclave and then buried. PRISM also treats wastewater from its wastewater treatment facility. This wastewater is treated using an effluent treatment plant. All waste collected is recorded on PRISM's customised software to keep track of and archive waste volumes for clients and PRISM operations.

Resources

Finances: PRISM's pilot program was funded by PRISM, the Canadian International Development Agency and the World Bank's Water and Sanitation Program. Through the success of the pilot program, PRISM received funding from the Government of Japan, through its Embassy in Bangladesh, to expand its program into six different districts.

The operations of PRISM's program are funded by the service charges paid by the healthcare establishments and the profits from selling shredded plastics to recyclers. For PRISM to expand its operations into new districts, increase transportation facilities or to upgrade waste treatment facilities, it requires external financial support from donor agencies.

People: PRISM consists of an executive committee with seven members. This committee makes organisational and financial decisions for PRISM and elects an executive director. PRISM employs 501 full-time staff, 152 of whom work for the medical waste management program.

Community support: The initial response from stakeholders to start the medical waste management program was very poor. PRISM conducted a series of

Medical waste being separated for correct disposal. PHOTOGRAPH: PRISM.

seminars and workshops with stakeholders and community members to gain support. Many community members and stakeholders have been very supportive of PRISM's program and the positive impact it has had in safely managing infectious medical waste.

Equipment: The medical waste management program requires covered trucks to efficiently service multiple healthcare establishments daily, as well as a shed to segregate the medical waste. The treatment of non-recyclable waste requires an incinerator and shed, an autoclave (sterilisation machine) and shed and an effluent treatment plant. Recyclable waste requires a storage area. An office is required for administrative operations.

Environmental and social benefits

Each day PRISM's program collects around 16 800 kg of infectious waste, 1050 kg of recyclable waste, 630 kg of sharps waste, 2100 kg of pharmaceutical waste and 420 kg of body tissue waste. The reduction of medical waste discarded into household waste roadside bins in communities has resulted in odour-free, clean air, a reduction in waste and soil contamination events and a subsequent reduction in health problems. PRISM's program has also provided employment opportunities to members of the communities in which it operates.

Barriers to success

Lack of available land: PRISM aims to have its medical waste management program established across the whole of Bangladesh. However, PRISM is constrained in establishing its program in other regions of Bangladesh due to a lack of available land to establish a new medical waste treatment facility. Without building nearby treatment facilities, transporting waste from regions of Bangladesh is financially unfeasible.

Attitude of service recipients: Obtaining support for and acceptance of the medical waste management program is difficult. PRISM conducts a series of meetings and workshops with healthcare establishments and government stakeholders to build support for its waste management service and to explain the benefits it can bring to the community.

Lack of government policy and coordination: The lack of government policies setting rules for how medical waste is discarded has made it difficult for PRISM's program to expand further. The lack of policies mean healthcare establishments are not required to discard their medical waste to designated treatment facilities. Hence, PRISM needs to put in considerable effort to convince healthcare

A team of medical waste collectors. PHOTOGRAPH: PRISM.

establishments of the social and environmental benefits of paying for PRISM's waste management service to manage medical waste safely and appropriately.

The lack of inter-ministerial coordination in the Bangladesh Government also makes it difficult for PRISM to pressure governments to put such medical waste policies in place.

Scalability and future outlook

It is PRISM's dream to operationalise the medical waste management program across all of Bangladesh. Serving all healthcare establishments in Bangladesh will vastly improve the situation of Bangladesh waste management streams and reduce health problems caused by mismanaged medical waste. PRISM is approaching donor agencies, governments and stakeholders in new districts of Bangladesh to expand its program.

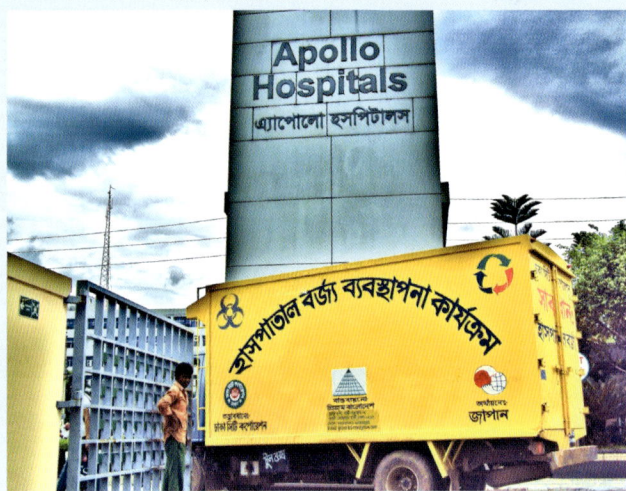

A truck collecting medical waste from a hospital to be taken to a recycling facility.
PHOTOGRAPH: PRISM.

Saahas

WASTE MANAGEMENT | RECYCLING | AWARENESS CAMPAIGNS

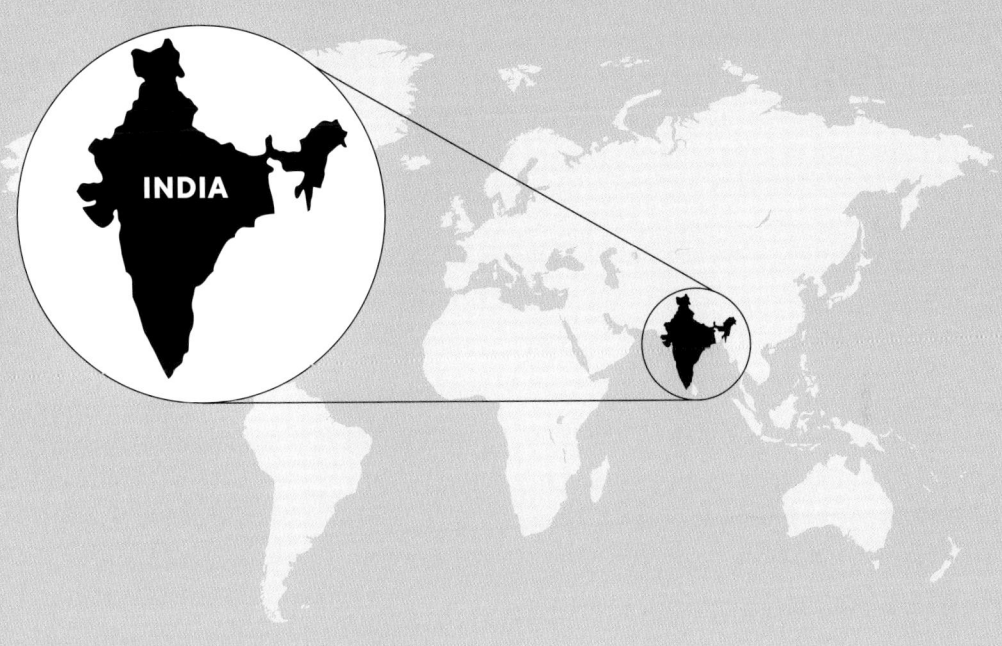

INDIA

| COLLECTOR | → | SORTER | → | RECYCLER/REPURPOSER |

Approximately
1200 tonnes of waste
per month

Set up 12 months,
operating since 2018

Multiple
municipalities

CONTACT
W: https://saahas.org/ • S: @saahas_ngo • E: info@saahas.org

SAAHAS IS A NON-FOR-PROFIT ORGANISATION THAT STARTED IN 2001 IN Bangalore in Karnataka, a state in the south of India. Its mission is to make India a leader of circular economy principles where nothing is wasted. Saahas specialises in waste management practices, awareness and activism, closely working with communities, administrators, businesses and policy makers. Saahas pilots and develops a range of innovative waste and resource management programs to implement at a municipality level.

One such program is the Swachha Udupi Mission, which implements source segregation of waste to facilitate resource recovery and reduce waste discarded into landfill.

> 'We all need to take ownership of our waste; we cannot dump our waste in other people's backyards and force them to bear the burden of our reckless consumption-based lifestyles.' – **SAAHAS**

The Swachha Udupi Mission program started in December 2018 and is a collaboration between Saahas, the Udupi City Municipal Council and self-help groups within Udupi City. The self-help groups typically consist of 5–15 members, usually women, who contribute financially to a pool of funds, which are then available for members to borrow. Prior to the program starting, the Udupi City Municipal Council and self-help groups were working together to collect waste from households around Udupi City. However, the household waste collected was mixed and could only be discarded into landfill. In addition, people were burning and dumping their waste in public areas.

Unloading of wet waste into a composting shed. PHOTOGRAPH: SAAHAS.

The Udupi City Municipal Council recognised that this was an unsustainable way to manage waste and that many components of waste could be used as an economically viable resource. For example, many types of plastic could be recycled, and organic materials could be turned into compost. At around the same time, Saahas was approached by HDB Financial Services, a non-banking financial company, wanting to fund a corporate social responsibility project in Udupi City. Partnering with the Municipal Corporation and self-help groups, Saahas used the funding provided by HDB Financial Services to start the Swachha Udupi Mission.

In its first year, the program was provided in eight municipalities around Udupi City; in its second year it was expanded to the entire 35 municipalities within Udupi City (an area of ~68 km²). At the beginning of the program, Saahas trained municipal corporation staff and members of self-help groups through various capacity-building sessions in matters such as waste collection, home composting, segregation and communication to households.

Door-to-door source segregation awareness campaigns were conducted by Saahas, municipal corporation staff and members of the self-help groups. Awareness campaigns were also held in apartments, schools and public venues every month.

The awareness campaigns focused on households starting their own home compost system and separating their own waste into three streams: wet/organic waste; dry waste, including materials such as plastic, paper, glass, cloth and metal; and domestic hazardous waste, including items such as sanitary and medical waste.

How the program works

Using vehicles provided by the Municipal Corporation, members of the self-help groups collect waste from households and shops. All household waste is collected, but only if it is separated. Households and shops are required to pay a fixed fee for this service.

Per month, the program collects around 400 000 kg of dry waste (which includes 16 000 kg of glass and e-waste), 800 000 kg of organic waste (including 60 000 kg of slaughterhouse waste) and 30 000 kg of domestic hazardous waste. The collected waste is then transported to processing sheds. Organic waste is shredded and composted. Any excess wet waste is sent to a biogas plant to generate energy. Domestic hazardous waste is given to designated processing facilities or is landfilled, and dry waste is sorted into different material types.

The recyclable, sorted, dry waste is sold to scrap dealers at variable market prices and the non-recyclable waste is sent to cement factories for coprocessing. Recyclable

SOURCE SEGREGATION OF WASTE

Segregated waste
collected from households

Organic waste
is composted

Domestic hazardous
waste sent to landfill

Dry waste sorted
into recyclables and
non-recyclables

Excess organic waste
sent to biogas plant

Recyclables
sold to local
scrap dealers

Non-recyclables sent
to cement factory for
co-processing

An overview of Saahas operations.

materials are sold to scrap dealers on a regular basis, and are sold to the dealer who offers the best market price.

Resources

Finances: The Swachha Udupi Mission was a corporate social responsibility initiative of HDB Financial Services. HDB Financial Services financed the first three years of the program, but it was Saahas that created the partnerships and training needed for the program to succeed. Other than the initial funding provided by HDB Financial Services, there is no continual external investment in the program.

The revenue from the sale of recyclable items and user collection fees goes to the self-help group members only. The project (and Saahas in general) does not benefit financially from the operations. The project is completely funded by HDB Financial Services.

People: Saahas has a 14-member team to facilitate, coordinate and manage the project. At the end of the initial three years, Saahas began training the municipality members and self-help group members to become self-sustainable. The initial project was only to set up the source segregation and processing centres in Udupi. Following the setup of the centres, the project will be officially closed and the responsibility of maintaining operations handed over to the Udupi City Municipal Council.

Equipment: The program operates using two sheds for windrow composting of organic waste, three sheds for sorting dry waste, an office space for administrative

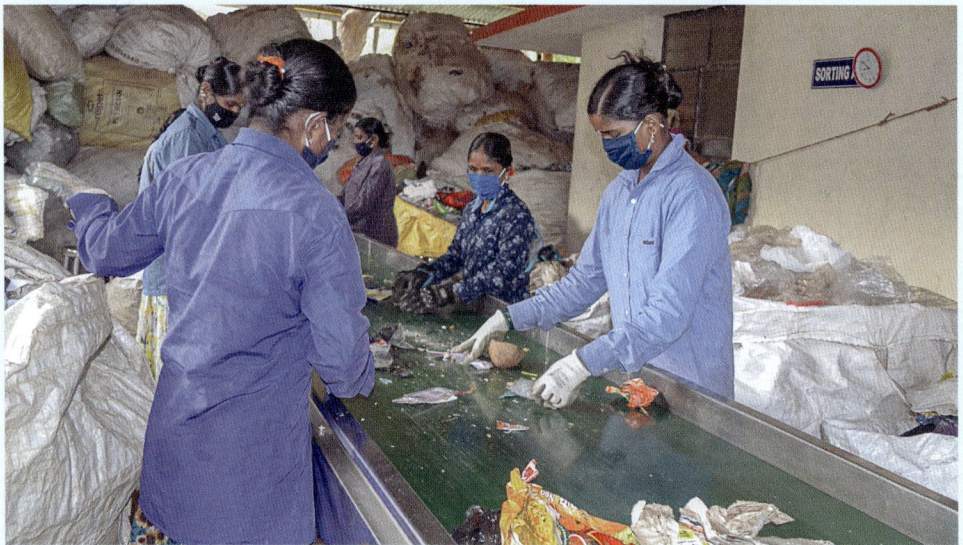

A conveyer belt used to sort waste types. PHOTOGRAPH: SAAHAS.

tasks, a biogas plant to generate energy from excess organic waste and vehicles to collect and transport household waste. The Udupi City Municipal Council provided sheds and vehicles to the program. Saahas has donated conveyor belts to improve the sorting efficiency of dry waste, as well as a compost turner and shredders to increase the composting process of organic waste. It has also installed floating trash barriers within the Udupi City Municipality to prevent waste from entering the ocean.

Environmental benefits

Open dumping of waste in Udupi City has been reduced by 50%, as has the frequency of people burning their waste. The awareness and education campaigns have increased the number of households starting their own home composting system and given households a sense of responsibility for the waste they generate. Households also take pride in contributing to a cleaner environment, and segregating waste at the source has diverted substantial volumes of waste from being discarded into landfill.

Social benefits

The Swachha Udupi Mission has provided a source of employment and income for many self-help groups. The members of these groups are predominantly women from disadvantaged sectors of the community. Self-help groups enable their members to collectively become socially and economically strong by providing employment and finances to all members.

Saahas and the program partners provide free dry waste bags and pamphlets to community members to encourage and ease the transition to behaviours that make segregating waste at the source easy. Personal protective equipment is provided to all self-help group members and Udupi City Municipality Corporation staff involved in the program.

During the capacity-building sessions provided by Saahas, hygiene kits were distributed to the Municipality Corporation and register books were distributed to the self-help groups to record data on waste volumes collected.

Saahas also creates green living spaces by converting open dumping spaces to small community gardens.

Barriers to success

Infrastructure: One of the biggest barriers the Swachha Udupi Mission has needed to overcome is the lack of processing facilities to sort and store dry waste materials.

The lack of processing space has restricted how much waste the program can collect and how long the sorted materials can be stored before sale to a scrap dealer. Due to the limited storage space, sorted recyclable materials need to be sold to scrap dealers on a regular basis.

The market price of recyclable materials changes frequently, making it difficult for program workers to stay up to date with current prices. This means workers are not always offered the best price for recyclable materials. However, because there is no space to store the materials, the workers must sell them and receive the lower rate. Saahas is currently trying to increase the space available for sorting and storing materials.

Community education: Another barrier the Swachha Udupi Mission has faced is getting households to segregate waste. Some households feel they do not need to sort their waste because they are paying for a waste collection service and that the service should include segregation. Continual awareness and education campaigns in the community help, as does training workers to have conversations with households on the importance of segregating waste at the source.

'If we keep dumping waste into landfills, then there won't be any land left for future generations.' – **SELF-HELP GROUP MEMBER IN UDUPI CITY**

Technology: Finally, for the program to grow and continue, better tracking of waste flows is required. Currently all data on waste volumes are recorded manually.

Roadside waste collection. PHOTOGRAPH: SAAHAS.

Saahas is looking at finding or developing software or apps to store, monitor, track and record data easily.

Scalability and future outlook

Programs such as the Swachha Udupi Mission have been implemented by Saahas in many municipalities across seven regions of India (Gurugram, Surat, Hyderabad, Ballari, Hubballi, Chennai and Bangalore). New programs require external investment, such as corporate social responsibility initiatives or grants, but the ultimate goal of Saahas is to create programs that can grow and become self-sustainable systems.

The first year of the Udupi program only serviced eight municipalities within the city. Starting the program small allowed Saahas to facilitate partnerships and systems between households, self-help groups and municipal corporations. Keeping the program small at the beginning also enabled problems to be resolved quickly and deliver awareness campaigns to all households in a timely manner.

Sevanatha Urban Resource Centre

WASTE MANAGEMENT | COMMUNITY BASED | EDUCATION

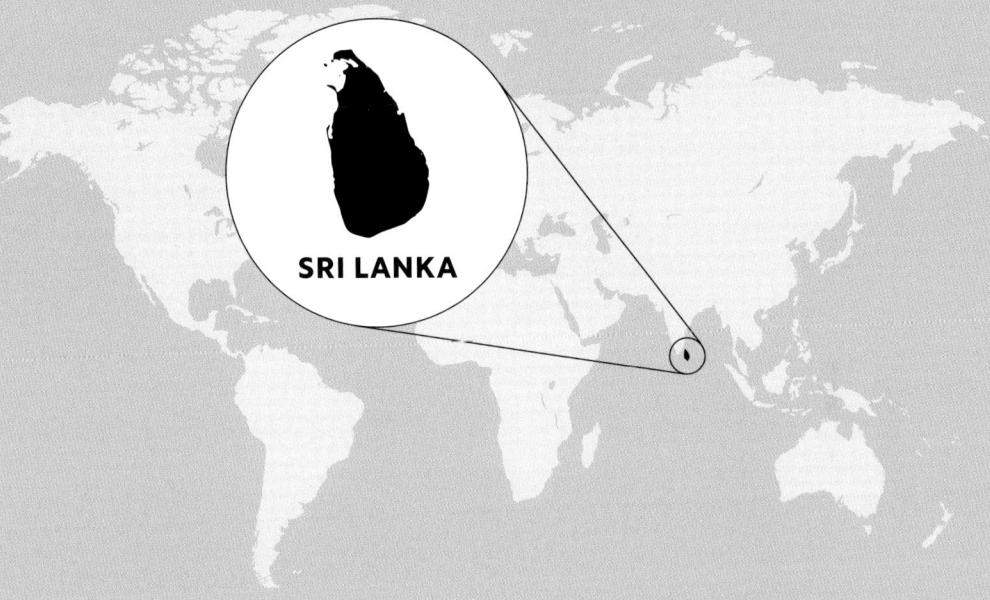

SRI LANKA

BROKER

Municipal Waste Recycling Project: 350 tonnes of waste per month and 890 kg of plastic waste per month Integrated Resource Recovery Centre: 390 tonnes of organic waste per month

Municipal Waste Recycling Project: Set up 1 year, operating since 2018 Integrated Resource Recovery Centre: Set up 1–4 years, operating since 2011

Municipality

CONTACT

W: http://www.sevanatha.org.lk/ ● S: @sevanathaUrbanResourceCentre
E: sevanata@sltnet.lk

THE SEVANATHA URBAN RESOURCE CENTRE IS A NON-GOVERNMENTAL organisation based in Colombo, Sri Lanka. Sevanatha was first established in 1989 as a grassroots activist organisation that improves the development of the urban community in Sri Lanka. Its mission is to be an agent for transforming the lives of those in urban and rural poor communities for them to become self-reliant and empowered members of society. Sevanatha predominantly works as a consultative body developing, trialling and promoting several community-based programs to revitalise and enhance the capacity of poor communities.

One way Sevanatha transforms the lives of poor communities is through the development and establishment of sustainable, decentralised waste management initiatives. Sevanatha recognised there were large amounts of ground and water pollution from haphazard waste disposal by people not acting responsibly and discarding their waste inappropriately. Sevanatha saw that if this waste could be managed at a local level, it could become a resource that benefits, rather than harms, communities.

In 2000, Sevanatha implemented its first community-based program to address waste in municipalities around Sri Lanka. This project developed, manufactured and trialled home composting bins, running community education events on the benefits of, and encouraging households to start, managing their own organic waste.

The initial compost bin program operated at a municipal level and successfully provided 40 000 bins to households. After this initial success, the program was continued by local entrepreneurs from the informal sector who manufactured the bins as a business and received support from their respective municipal governments.

The success of the compost bin program encouraged Sevanatha to continue finding local solutions to waste management issues that improved the livelihoods of poor communities in Sri Lanka. The following describes two projects established by Sevanatha: an Integrated Resource Recovery Centre in Ratnapura and Matale municipalities and a Municipal Waste Recycling Project in the Dehiwala–Mount Lavinia municipality.

How the programs work

Integrated Resource Recovery Centre

In 2007, Sevanatha initiated the Integrated Resource Recovery Centre, a municipal program that incorporates the reduce, reuse and recycle principles into a business model for urban solid waste management. The aim of the program was to promote the benefits of a clean environment and generate sufficient income from urban waste to operate a sustainable municipal waste management system.

A beach-side bottle collection station. PHOTOGRAPH: SEVANATHA.

The program involves the collection and recovery of organic and recyclable materials from within the municipality. These materials are delivered to the centre, where organic materials are composted and plastic and paper materials are shredded and baled.

The program was first piloted in the town of Matale, with the success of the pilot enabling Sevanatha to receive financial support from the United Nations (UN) Economic and Social Commission for Asia and the Pacific (ESCAP) and Waste Concern Consultants, Bangladesh, to expand the project and construct two integrated resource recovery centres, one in Matale in 2011 and the other in Ratnapura in 2014.

The project expansion was also supported by the local governments in both towns. The continuing operations of both centres are currently managed by Micro Enrich Compost Pvt Ltd, a subsidiary company established by Sevanatha in partnership with the Ratnapura and Matale municipal governments to promote public–private partnerships in the municipal solid waste management sector.

Sevanatha continues to support the integrated resource recovery centres by providing technical operations support and administrative assistance to Micro Enrich Compost. Sevanatha also promotes and mobilises community members to separate their waste and to provide it to the collectors of the integrated resource recovery centres.

Municipal Waste Recycling Project

In 2018, with the support of the United States Agency for International Development (USAID), Sevanatha started the Municipal Waste Recycling Project in the Dehiwala–Mount Lavinia municipal area. The program was initiated to address the large volume of waste generated by the municipality: almost 1 kg of waste per person per day (242 000 kg of waste per day from a population of 250 000 people). Due to deficiencies in the municipality's existing waste management system, a considerable amount of this daily waste leaked into the environment.

The project was established to reduce the quantity of plastic waste entering the environment by incorporating independent waste collectors from the informal sector to help improve the municipality's waste management operations. Waste collectors receive organic and recyclable materials from households, selling recyclable materials to recyclers and delivering organic material to the municipal composting facility, which is operated as a partnership between Sevanatha and the municipal government.

In addition to waste collection by the informal sector, the project involved establishing 26 community collection points for polyethylene terephthalate (PET)

Community member using one of the PET plastic bottle community collection points established by the Sevanatha Recycling Project. PHOTOGRAPH: SEVANATHA.

bottles, the creation and distribution of over 1000 reusable cloth shopping bags to community members in an effort to reduce plastic shopping bag use and litter and the installation of two waste traps that capture waste floating down waterways before it enters the marine environment.

The project also involved Sevanatha running a large community training, awareness and education campaign to promote and mobilise community members to: become waste collectors, understand the benefits of recycling, reduce the disposal of waste into canals within the municipality, raise awareness of the impacts of pollution and separate their household waste. The campaign was run as a series of workshops hosted at religious centres, schools and community centres, as well as community beach clean-up events.

Resources

Finances: Sevanatha acquires funding from a range of internal and external sources. Sevanatha self-funds projects at the initial development and trial stages. To fund the commencement of pilot projects, Sevanatha has approached and received financial support from development agencies, such as the UN Human Settlements Programme (UN-Habitat), World Vision, UN ESCAP, the UN Development Programme, the USAID and the World Bank.

Funding for the continuation of programs after their initial pilot stage commonly comes from a combination of the program's profits and funding from associated municipal governments or private business investments.

People: Sevanatha's projects and consultancy assignments are run by six senior management staff, who oversee 20 operational and project staff and two support staff. Additional field staff are employed to undertake short-term field activities (normally <6 months). The staff implement programs, provide technical support and advice, and acquire funding for new and continuing projects.

> 'Our principle aim is to act locally with available resources and make a big impact with support from organisations like local government and development agencies.' – **KA JAYARATNE, PRESIDENT, SEVANATHA URBAN RESOURCE CENTRE**

Environmental and social benefits

Integrated Resource Recovery Centres

Both the Matale and Ratnapura centres generate profits from the compost they produce, which is sold under the Micro Enrich brand, and by selling collected recyclable materials to local recyclers. The Matale and Ratnapura centres are

A Sevanatha-organised beach clean-up. PHOTOGRAPH: SEVANATHA.

currently handling 9 and 5 tonnes of organic waste a day respectively. Both centres have also provided employment and improved the livelihoods of low-income community members, particularly women, who either collect material or work at the centre.

The program has also paved the way for other municipalities in Sri Lanka to consider implementing a public–private partnership business into their urban waste management system.

Municipal Waste Recycling Project

Eighteen months after the project started, segregation at source and recycling behaviours have been established in 443 households within the municipality. Over 6300 tonnes of household waste has been collected from the community, and 16 tonnes of plastic waste has been recovered and either recycled, reused, exported or incinerated.

By mobilising and incorporating the informal sector into the municipality's waste management system, the municipality was able to increase the efficiency and capacity of its waste management and to recover significantly more waste from the community to reduce the amount of waste entering landfill and the environment.

During the 18 months of the project's operations, waste recovered by the informal sector increased from an average of 3 to 10 tonnes per month. The informal sector also collected over 6 tonnes of plastic waste, almost matching the 9 tonnes collected by the municipality staff.

Over 500 community members attended the training workshops, which resulted in increased women participation and employment in community waste management. The recovery of waste materials, the promotion of correct waste disposal behaviours and the establishment of decentralised waste management programs have resulted in cleaner drains, canals, beaches and public spaces in the serviced areas.

In addition, throughout the project, Sevanatha has promoted informal waste collection as a respectable livelihood in communities. Many local communities now see waste as a resource and waste collection as an additional source of income for many low-income families. For example, many women from low-income communities have become waste collectors, recovering valuable waste materials in their local area as a source of income.

Barriers to success

Acquiring funding: The biggest challenge Sevanatha faces with any project is acquiring funding for the continuation of the project after the seed funding has been used up. In some cases, the projects Sevanatha implements do not become self-sustainable after the initial funding has ended. For example, only 25% of the operational costs for the integrated resource recovery centres were covered from the sale of compost and recycled materials.

For the centres to continue operating, Sevanatha negotiated with the municipal governments to finance the operations of the centres while Sevanatha provided technical support. Sevanatha overcomes these funding constraints by involving municipal governments at the beginning of any new project and incorporating public–private partnerships into their project designs.

Municipality support: Getting Sri Lankan municipalities to recognise and acknowledge the benefits of incorporating the informal sector into decentralised waste management systems remains challenging. Centralised waste management systems are still prominent in many municipalities around Sri Lanka, and there is a lack of effort from many governments to change to a decentralised system. Sevanatha continues to promote the success of waste management systems that incorporate the informal sector and advocates waste collection as a respectable livelihood.

Good communication with society and local government is needed to run a successful project. Building relationships between government institutions and urban local authorities has allowed for the efficient performance of Sevanatha's activities within the targeted communities.

Scalability and future outlook

When Sevanatha first started, operations were limited to four cities: Colombo, Dehiwala–Mount Lavinia, Kandy and Galle. As the success of Sevanatha's programs grew, operations expanded during the 1990s to over 13 districts in six provinces.

At the beginning of each new project, Sevanatha trials the program at a small scale before implementing larger pilot programs or expanding the programs. Starting small enables Sevanatha to acquire resources to expand the projects and to quickly solve any problems that arise, without the potential for large consequences on the continuation or success of the program.

Over the past 20 years of operation, Sevanatha has built a strong positive reputation of trust within the communities it works in and Sri Lankan society more broadly. Sevanatha continues to work closely with local governments, informal waste sectors and disadvantaged communities. Building a strong trust relationship with communities and governments at the small, local level, has allowed Sevanatha's programs to be established in communities right around Sri Lanka.

Building trust and inclusions between all stakeholders is key to the success of Sevanatha's programs and allows Sevanatha to continue trialling and establishing programs that improve the management of waste around Sri Lanka. Sevanatha is currently piloting more programs with communities and local governments to find solutions to the problem of urban waste and plastic pollution.

Solid Waste Collection and Handling (SWaCH) Pune Cooperative

COMMUNITY BASED | WASTE COLLECTION | RECYCLING

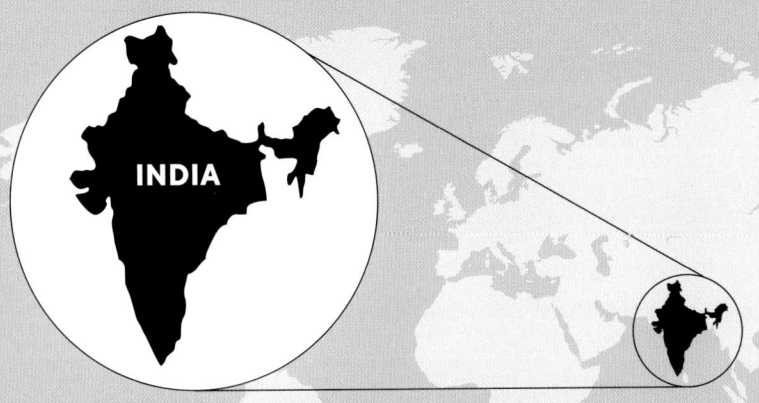

INDIA

COLLECTOR → SORTER → RECYCLER

1400 tonnes of municipal solid waste per day

Set up 2 years, operating since 2007

Multiple municipalities

CONTACT

W: https://swachcoop.com/ • W: www.wastepickerscollective.org

E: swachcoop@gmail.com • T: +91 9765 999 500

LONG AGO, IN 1990, WASTE PICKERS IN THE PUNE AND PIMPRI CHINCHWAD municipal districts in India were treated like the trash they collected. People would shoo them away like a stray dog and cover their noses when they passed. This was hurtful to the waste pickers. They were not sorting through rubbish because they enjoyed it; they were there to collect recyclable material so they could sell it and use the money to feed themselves and their families.

One summer's day in 1993, a group of waste pickers came together to try and change public perceptions and make waste picking a dignified profession. These waste pickers formed Kagad Kach Patra Kashtakari Panchayat (KKPKP), a membership-based trade union that asserts the status of waste picking as a working profession and the crucial role of waste pickers in city waste management and environmental sustainability.

> 'We were there because we wanted the recyclables ... Our mothers and grandmothers had done it before us. It was this work that brought us money to feed ourselves and our families, so we did it.'
> 'That was the turning point for us. That day we decided to stand tall. That day we decided to walk with dignity on the road we had never travelled.' — **MEMBER OF KAGAD KACH PATRA KASHTAKARI PANCHAYAT**

Through case studies completed by KKPKP that quantified the contribution waste picker operations made to the recovery of waste and the money these informal operations save the Pune municipality, KKPKP was able to push for better working conditions and the integration of waste pickers into formalised waste collection and management systems. Taking advantage of new municipal solid waste laws that required the segregation of waste and door-to-door waste collection services, KKPKP launched a pilot program in 2005 with the Pune Municipal Corporation in which waste picker operations were incorporated into door-to-door waste collection.

In 2007, two successful years of the pilot scheme led to the formation of a formal, workers-owned cooperative between waste pickers and the Pune Municipality Corporation called the Solid Waste Collection and Handling Pune Cooperative (SWaCH). This was the first cooperative of its kind in India. The SWaCH waste collection program involves members providing household door-to-door waste collection services and the Pune Municipal Corporation providing members with infrastructure and management support for program operations.

How the program works

In Pune, households can dump their waste into roadside dumpsters provided and emptied by the Pune Municipal Corporation or into vehicles (trucks or carts) in public spaces. This dumpster service is covered by a municipal tax that all households are required to pay. To receive the door-to-door SWaCH service, residents must pay a small additional fee. Households segregate their waste into wet (organic) waste, dry waste (recyclables, general waste) and sanitary waste. SWaCH waste pickers collect the waste using carts and deliver the waste to trucks, which transport the waste to sorting sheds.

At the sorting sheds, members further sort the waste and recycle it into another 40 different material streams. Organic waste is sent to composting units or biomethane power plants. Recyclable materials are sent to a warehouse to be shredded and baled. This baled material is then sold on to recyclers at market price. Non-recyclable material is transported to municipal dry waste processing plants of refuse-derived fuel plants.

SWaCH also runs awareness campaigns and rallies to maintain support for the waste collection services and for the segregation of waste by households.

Resources

Finances: The pilot program was implemented in collaboration with the Department of Adult Education and the Shreemati Nathibai Damodar Thackersey Women's University. SWaCH formed in 2007 and was self-sustaining by 2014–15. The SWaCH waste collection program continues to operate from the service fee paid by households who have their waste collected and from revenue generated by selling material to recyclers.

Through the partnership with the municipal corporation, SWaCH waste pickers have ensured rights to collect the high-value recyclable material from households. SWaCH also facilitates its members' access to government-sponsored social

Two of the many SWaCH professional waste pickers.
PHOTOGRAPH: SWACH.

welfare benefits, such as subsidised healthcare, insurance and scholarships for children.

People: The pilot program involved 1500 waste pickers servicing 125 000 households in Pune. Since SWaCH formed, both the service area and the number of members have expanded. SWaCH has over 3700 waste picker members servicing over 880 000 properties with door-to-door waste collection. SWaCH also include members who run the sorting sheds, composting and baling facilities, and those who run awareness and education campaigns at schools and community centres.

Community support: SWaCH easily received community support when it started because it was a partnership between the KKPKP union and the municipal corporation. Most members of SWaCH are also members of KKPKP, but not all KKPKP members have joined SWaCH. To generate support in Pune communities to pay for a door-to-door waste service fee, segregate waste at their homes and acknowledge waste picker contributions to the community, SWaCH runs awareness campaigns and rallies with school students, citizen groups and elected representatives.

Equipment: Most of the infrastructure and equipment required for program operations are provided by the Pune Municipal Corporation. SWaCH provide uniforms, identification cards and personal protective equipment to each waste picker. Over 2000 pushcarts are used to collect waste from households. Almost 100 waste pickers have upgraded (via private loans) to small, motorised vehicles to provide waste collection services.

This waste is transferred to small (300-kg capacity) trucks, which transport the waste to sorting sheds. SWaCH currently uses 64 sheds and 140 portable

A SWaCH collection trolley with dividers to segregate waste according to type.
PHOTOGRAPH: SWACH.

material recovery facilities to sort the waste. Sorted waste is then transferred in large (1000-kg capacity) trucks to warehouses for further sorting and baling. Finally, very large (8000-kg capacity) trucks are used to transfer baled material to recyclers. Carts, trucks, sorting sheds and warehouses are provided by the Pune Municipal Corporation.

Environmental benefits

SWaCH's waste collection service covers households in urban centres, slums and fringe villages in Pune. SWaCH collects around 1400 tonnes of household solid waste per day. This waste includes around 220 tonnes of recyclable material (e.g. plastic, paper, metal and glass) that is diverted to recycling facilities and 45 tonnes of sanitary waste collected through SWaCH's Red Dot campaign.

Prior to SWaCH's collection service, houschold waste was dumped along streets. This dumping made the streets very smelly, and it was very unhygienic for waste pickers to sort through the waste to find valuable material. Since SWaCH's program, there is very little waste dumping along the streets of Pune, and waste pickers can safely and hygienically sort through segregated waste collected from households.

Social benefits

Municipality savings: Each year, the SWaCH program and partnership save the Pune Municipal Corporation over US$13.6 million. By incorporating waste pickers into the formal door-to-door waste collection service, the Pune Municipal Corporation saves on additional administration and operational costs.

Employing women and social welfare: Around 75% of SWaCH members are women and 80% of KKPKP members are women. Most women are from marginalised, disadvantage castes within Pune. SWaCH alleviates poverty in Pune by ensuring all their members are paid equally from the household service fee.

When waste pickers become a SWaCH member, they are given access to a social welfare scheme that provides health and life insurance and a contributory pension. In addition, SWaCH applies for grants to support their members with providing education for their children or obtaining interest-free loans for household necessities.

SWaCH Plus: In addition to door-to-door waste collection service, SWaCH now provides a SWaCH Plus program that offers products, programs and services that involve citizens in recycling and sustainable living, such as composting services (5 tonnes per day), as well as reuse and recycle collection events for old electronic items, furniture, bicycles and clothing (>25 tonnes per month). The items are reused or resold to waste pickers and the urban poor.

Red Dot awareness campaign: To improve the safety and health of its members, SWaCH promotes the safe disposal of sanitary waste. Every day SWaCH waste pickers handle 20 tonnes of household sanitary waste, some of which is unwrapped, exposing the waste pickers to harmful pathogens. Through awareness campaigns and instructional videos, SWaCH has educated households on the correct disposal of sanitary waste and to mark this waste with a red dot so that it can be easily identified as sanitary waste at the time of collection. The Pune Municipal Corporation is in the process of setting up India's first mechanical recycling plant for sanitary waste based on SWaCH's segregated waste collection system.

Barriers to success

Infrastructure: Currently SWaCH does not have adequate shed and warehouse space to sort, bale and store the collected household waste. Through negotiations, the Pune Municipal Corporation is building more storage sheds.

User fee and waste segregation enforcement: A constraint on the efficiency and success of SWaCH is getting residents to pay the monthly fee for their waste to be collected. Some households are slow to pay the service fee, which, in turn, means waste pickers do not receive their pay. Some households do not appropriately segregate their waste, which creates more work for the waste pickers.

In some instances, SWaCH waste pickers will refuse to collect the unsegregated waste if it exposes them to health and safety hazards. SWaCH has members who will visit households to provide information and raise awareness as to the benefits of the SWaCH service, as well as to encourage households to use the service and correctly segregate their waste.

> 'The minute you mix waste, it gets contaminated. When dry waste and wet waste get mixed, you are spoiling the quality of the materials … this is why waste segregation at the source is so important.' — SUCHISMITA PAI, SWACH

Scalability and future outlook

SWaCH is aiming to expand its program beyond Pune Municipality. It has created a standard operating procedure so that other municipal corporations can identify and integrate waste pickers into formal municipal waste collection services. The standard operating procedure will include social welfare schemes for waste pickers, data management and the designs and operations of decentralised material recovery facilities.

SWaCH is also in the pilot stages of 100% *in situ* composting-based zero-waste wards. This pilot program involves all organic waste within Pune wards to be composted within the wards, whether in large windrow sheds or in home composting systems.

Finally, SWaCH has started extended producer responsibility-based recycling services for low-value plastic waste, such as soft packaging, and is already collecting over 100 tonnes per month.

SWaCH has more than 3700 waste picker members. PHOTOGRAPH: SWACH.

The Skilled Women Initiative

EMPOWERING WOMEN | UPCYCLING | EDUCATION

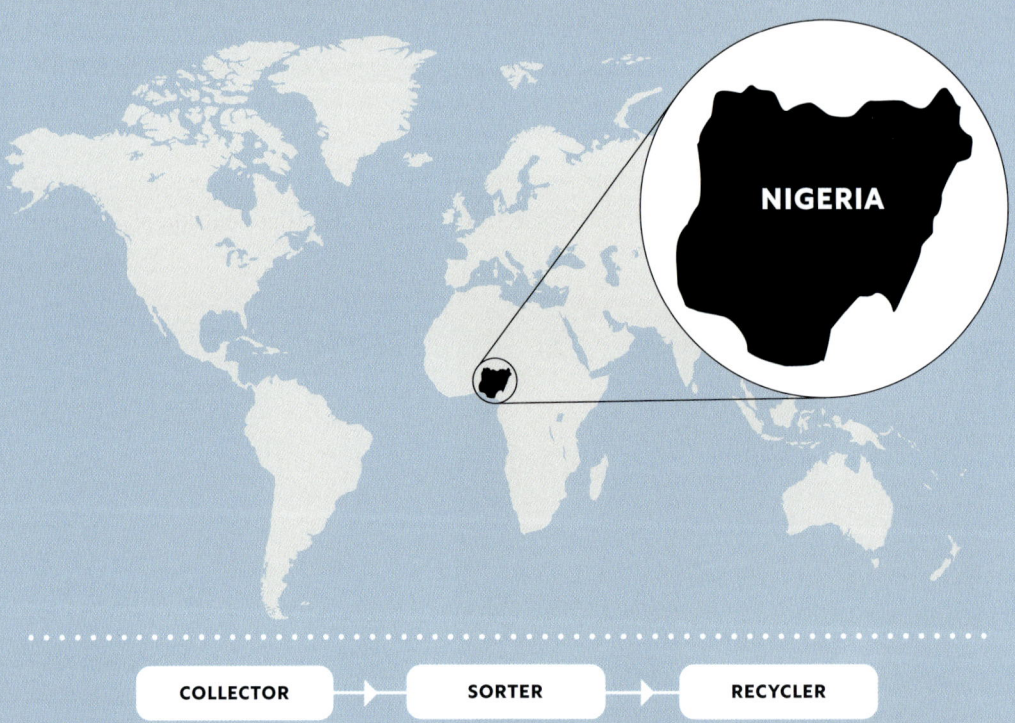

NIGERIA

COLLECTOR → SORTER → RECYCLER

173 kg of waste per week, including plastic, tin cans, glass jars, textiles and leather

Set up 2 years, operating since 2018

Multiple municipalities

CONTACT

W: www.tswini.org/ • S: @TSkilledwomen; @Theskilledwomen
L: https://www.linkedin.com/company/theskilledwomen/?originalSubdomain=ng
E: theskilledwomen1@gmail.com • E: hello@tswini.org • T: +23 480 9900 0502

THE SKILLED WOMEN INITIATIVE (TSWINI) STORY BEGINS WITH CHISOM Nwankwo, her three friends Chidiebere Obi, Iquo Offiong and Koyona Duke and three problems in their local community. These four women were overwhelmed by the volume of textile and leather waste the fashion industry produced, the amount of plastic and metal littered in the community and the lack of employment for refugee women and women displaced internally by insurgency.

Using their expertise in environmental protection and policy implementation that they gained from working together at Nigeria's Federal Ministry of Science and Technology, they decided to start an upcycling program that would not only reduce waste, but also create employment in the community.

In 2017, Chisom and her friends self-funded the first Skilled Women Initiative, a vocational skills centre that specialises in upcycling tin, textiles, leather, plastic and glass into a range of clothing and household products. TSWINI first started at the Kuchingoro Refugee and Internally Displaced People Camp but, due to complications, the centre was moved to the current location at Durumi Refugee and Internally Displaced People Camp.

The Durumi TSWINI centre has trained and employed over 500 refugee and internally displaced women and young people in the north central region of Nigeria.

Women learning to make products to sell.
PHOTOGRAPH: TSWINI.

How the program works

The TSWINI centre has five departments specialising in training participants in specific vocational skills. These departments are sewing and tailoring, upcycling and basket-coating technology, agricultural technology, household care products and photography and media. Participants learn six major skill sets at the training hub, including sewing, the production of household care products, bead work, plastic and fashion waste upcycling, basket weaving, photography and a simple introductory information technology course.

Each week the centre collects and processes 50 kg of textile offcuts from fashion designers, 50 kg of polyethylene terephthalate (PET) bottles, 50 kg of tin containers, 10 kg of leather offcuts from furniture makers and 13 kg of glass jars. The centre currently employs 65 (but has the capacity to employ 100) refugees and internally displaced persons, who are paid commissions on each product they make.

Each specially crafted product made by TSWINI is marketed under its 'Kwando' range. These products include liquid soap, scented candles, hand sanitiser, petroleum jelly, disinfectant, fabric bags and fashion accessories. All Kwando products are packaged in upcycled plastic bottles that are covered in a raffia woven basket. The word *kwando* in Hausa language, predominantly spoken in Niger and the northern states of Nigeria, translates to 'basket' in English.

Some of the many products produced by the TSWINI team. PHOTOGRAPH: TSWINI.

TSWINI has also started a new product line called IKPA fabric. IKPA upcycles textile waste into a woven fabric used to make fashion accessories and home decor pieces. Anyone can deliver their used clothes and fabrics to the TSWINI centre, where the textiles are sorted and woven into unique, colourful IKPA fabric.

Resources

Equipment: For the centre to operate, it requires a building where collected waste can be stored and sterilised, an area where the cleaned material can be upcycled into products and an area to store finished products for sale. TSWINI received 70 sewing machines as a donated from the Nigerian–German Centre for Jobs, Migration and Reintegration. This donation has also allowed the centre to increase the amount of material it can upcycle and the number of people it can train and employ.

Finances: From its conception, TSWINI has been developed as a self-sustaining system. The initial stages of TSWINI started out very small and gradually grew through word of mouth and social media. Within a year, the success of TSWINI's program was recognised and it received its first grant of US$2000 from a national bank in Nigeria. This funding enabled the centre to purchase more equipment and grow so the centre could upcycle more material and train more people.

The continuation of the centre relies on and is sustained by profits from selling the upcycled Kwando and IKPA product lines. TSWINI has also received financial support from the Nigerian National Commission for Refugees, Migrants and Internally Displaced Persons.

Community support: During the COVID-19 pandemic, TSWINI started a project and campaign called #facemaskbyIDPs and trained TSWINI participants to make reusable cloth face masks. Participants made over 1000 face masks for themselves, their families and the community. Face masks were also sold to generate income for food and basic necessities.

Environmental and social benefits

In addition to training and providing employment for over 500 refugees and internally displaced persons, TSWINI has returned over US$7000 to the community and has significantly reduced textile and plastic waste. TSWINI have upcycled over 500 kg of textile waste. Within the target communities TSWINI services, predominantly the states of Abuja and Nasarawa in Nigeria and Niger, there has been a 30% decrease in litter over the past 3 years.

'Our mission is Learn, Earn, Blend ... At TSWINI we are making sure that our beneficiaries get access to learn new vocational and tech skills of their choice, earn money from the products they make, and eventually leave the camps and blend back into their local communities.' – **TSWINI**

Community integration: TSWINI's aim is for every women to have the entrepreneurial skill set and resources to pursue employment outside of the initiative after 18 months of training. TSWINI designed an integration program where participants train in a skill of their choice, open a savings bank account, sell products under the Kwando brand and save 50% of the sale proceeds over an 18-month period. At the end of the 18 months, TSWINI will match the profits saved by each participant, and these funds will be used by participants after they leave the camp to start a small business and to cover upkeep costs.

In July 2020, 100 TSWINI participants signed up for the TSWINI integration program. In addition, TSWINI is investigating options to provide post-camp housing and healthcare facilities for leaving participants.

The skilled fingers of TSWINI staff weaving products. PHOTOGRAPH: TSWINI.

Barriers to success

TSWINI operations and growth are constrained by the space available to store collected waste, financial costs to train more refugees and internally displaced persons, government regulations and policies on collecting, storing and repurposing solid waste and expanding the marketing capacity of the Kwando product line and upcycled clothing.

Scalability and future outlook

Starting small allowed TSWINI to trial and design different waste materials to products, find a suitable location to establish the centre, organise the sources and collection of waste materials and solve any problems that arose with few major consequences to the continuation of the program. Fixing problems while the program was small has enabled TSWINI to grow organically and create a self-sustaining system, from sourcing and collecting waste all the way to designing and marketing their upcycled Kwando product line.

TSWINI is now at the point where it can replicate the current system and centre at Durumi Camp to other refugee and internally displaced persons camps around Nigeria. The TSWINI system is scalable and can start small while equipment, training, waste material collections, product marketing and sales are established.

Sports and Skills Hub: More than 40% of young people at the Durumi Camp are boys. Due to the success of TSWINI's women-focused training hub, TSWINI has started a mentoring and training hub for boys at the camp so they also gain vocational skills that enable them to eventually leave the camp. The Sports and Skills Hub will provide hands-on training in carpentry, upcycling of tyre waste, agricultural technology (including aquaponics, housing poultry and greenhouse construction), fashion technology, photography and information technology. The Hub will also incorporate sports as an incentive to bring boys to the Hub and build teamwork skills.

A TSWINI facility with team members making products to sell. PHOTOGRAPH: TSWINI.

TierraMar: SeaNet Indonesia

REDUCING OCEAN POLLUTION | RECYCLING | COMMUNITY BASED

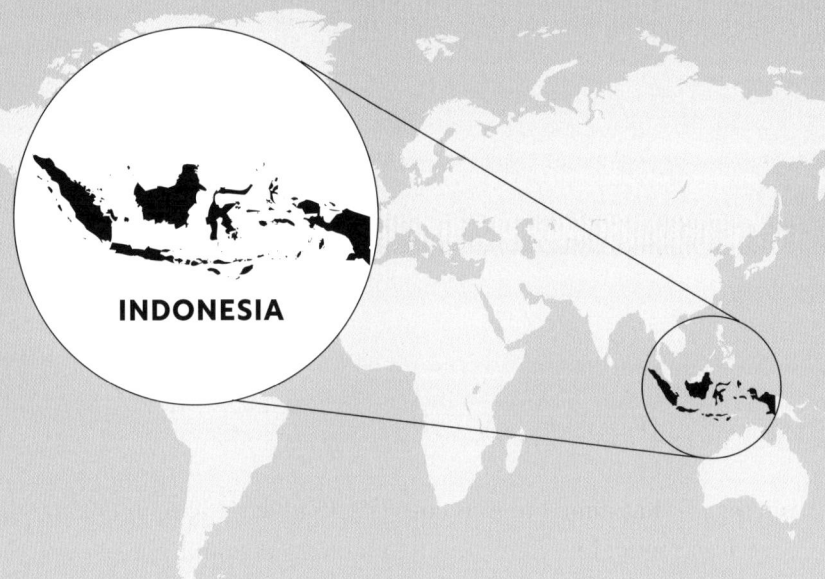

INDONESIA

COLLECTOR ➤ SORTER

18 tonnes of discarded nylon fishing nets

Set up 14 months, operated for 2 years; post-2019, nets are collected but no buyer has been established

Community wide

CONTACT

W: https://tierramar.com.au/portfolio-view/seanet-indonesia/ ● S: @TierraMar_Aus
L: https://www.linkedin.com/company/tierramar-consulting

SeaNet Indonesia community recovering fishing nets. PHOTOGRAPH: TIERRAMAR.

TIERRAMAR IS A NOT-FOR-PROFIT ENVIRONMENTAL ORGANISATION BASED IN Australia that provides strategic solutions for the conservation and sustainable management of biodiversity and natural resources. In 2017, TierraMar, in partnership with the Coral Triangle Center, started an 18-month pilot program within the Arafura Sea region that aimed to find cost-effective, scalable solutions to improve fisheries practices, reduce by-catch, improve market access and increase compliance with fisheries regulations.

The overarching aim was to develop SeaNet Indonesia as a potential large-scale model of integrated fisheries extension services to alleviate poverty within the Coral Triangle region. SeaNet was implemented in communities in two regions of Indonesia where people's livelihoods depend on marine resources and fisheries: the Merauke District in Papua Province and the Southeast Maluku District in Maluku Province.

One scalable solution trialled during the SeaNet Indonesia program was to reduce the amount of fishing gear abandoned, lost or otherwise discarded to the marine environment, and to find a market solution to recover and recycle fishing gear already in the marine system within Merauke and Southeast Maluku.

How the program works

The recovery and recycling of fishing net waste, or ghost gear, officially started in October 2017 and was the first project in Indonesia to address ghost gear. TierraMar partnered with the Zoological Society of London to establish a recycling program based on the Zoological Society of London's Net-Works program in the Philippines.

The project was introduced into the remote communities of Merauke, whose livelihoods rely on barramundi and shark fisheries. TierraMar decided that if it could get the recycling program to work first in the most difficult remote community, then applying the program to other communities would be simpler.

When TierraMar first arrived in the Merauke communities, mountains of fishing nets occupied the coastlines and riverbanks. Some fishers in the community had previously sold nets to a buyer in Surabaya, Indonesia, who recycled the nets into plastic bags. However, this previous market for nets was not a viable solution to the ghost gear problem because the fishers received little money for each kilogram of net they sold, and the process did not involve or support the whole community.

The aim of the recycling program was to develop a community-driven initiative to recover at least 5 tonnes of nylon fishing net waste and export the nets to Aquafil's facility in Slovenia, where the nets would be recycled into carpet tiles under the EcoNyl brand.

The sale of nets would provide an additional viable source of income to the community and allow the community to reclaim their coastline from fishing debris. TierraMar established Aquafil as a buyer for the nets using the pre-existing business relationship between Aquafil and the Zoological Society of London's Net-Works program in the Philippines.

Before the nets could be shipped from the local port in Merauke to the international port in Surabaya and on to the destination port in Slovenia, community members needed to collect and thoroughly clean each net. The cleaned nets were then transported to a third party, who compacted, baled and stored the nets ready for shipment. To reduce shipping fees, all processed nets were shipped as one unit to Slovenia at the end of the year.

A SeaNet Indonesia team member weighing a recovered fishing net.
PHOTOGRAPH: TIERRAMAR.

In addition to the net recycling project, TierraMar worked with the local fishers to determine how their fishing practices could be altered to improve the quality of their catch, reduce by-catch and reduce the probability of their gear, particularly nets, becoming damaged or lost. Through discussions with the fishers, TierraMar ran a series of best-practice fishing workshops for multiple communities involved in the broader SeaNet Indonesia program.

Resources

Finances: The first year of the net recycling project was funded under the SeaNet Indonesia program, financed through the Australian Government's Coral Triangle Initiative support program. The continuation of the recycling project was funded for a second year by a grant received from World Animal Protection.

After the second year of funding, TierraMar aimed to have established the recycling project as a self-sustaining community program. However, due to exorbitant shipping export fees consuming most of the profits made from the sale of the fishing nets, the community could not fund the continuation of the project after external funding ended.

People: For each country that TierraMar operates in, there is an in-country manager who acts as a government liaison and provides technical support to each program within the country. Within each program, there is a program coordinator who runs the on-ground logistics and operations and manages officers in charge of specific components of the program, such as a fisheries extension officer or a women's empowerment officer. Partnerships are formed with local non-governmental organisations and governments to build their capacity and provide opportunities to ensure sustainability of ground outcomes with communities.

Community support: Before the SeaNet Indonesia program officially started, TierraMar visited communities within Merauke and Maluku to ask what the biggest challenges that they faced were, and what resources they required to overcome these challenges. These initial community discussions and project-scoping visits enabled TierraMar to establish trust with members of the community and gain community support for projects under the SeaNet Indonesia program.

Involving the community at the beginning of a project in its design ensured that the project was driven and owned by the community, not by TierraMar.

Environmental benefits

The community processed 10 tonnes of net in the first year and 8 tonnes in the second. Removing such large quantities of net enabled the community to reclaim

Recovered fishing nets being cleaned in preparation for recycling.
PHOTOGRAPH: TIERRAMAR.

previously inaccessible fishing grounds and coastline for recreational activities and increased the aesthetic value of the community.

Social benefits

A core element of projects within the SeaNet Indonesia program was to involve all stakeholders, from community residents up to the governor of the province (the *Bupati*), to improve fishing practices and alleviate poverty. SeaNet brought multiple government departments together who previously had little collaboration with one another and had little engagement with local communities.

TierraMar worked closely with government departments to build their collaborative and community engagement capacity to achieve collective goals and empower their communities. The capacity building and training alleviated tension and mistrust between communities and their government and established long-term collaborative and helpful relationships between both parties.

Barriers to success

Shipping costs and requirements: a major challenge of the net recycling project was the expense of exporting the nets from Merauke to Surabaya, then on to Slovenia. At the time of the project, no viable market had been identified to sell the nets to other than Aquafil. Shipping nets internationally is a very expensive process. In addition, although the fishing nets were cleaned to the high standards required by Aquafil, the nets were considered hazardous waste under international shipping export legislation and, as such, were burdened with additional shipping fees.

At the end of the 2-year program, TierraMar and the Merauke communities had a net recycling system that worked. However, the shipping fees made it financially prohibitive to establish a self-sustaining system, without a local buyer in the region.

Infrastructure: The first year the net recycling project operated, only one net compactor and baler were available to the community in Merauke. The compactor and baler were not suitably designed to handle fishing net waste and unfortunately the machinery broke.

In the second year, TierraMar helped the community use the funds from the first-year sales of fishing nets to purchase a new compactor and baler built for handling fishing nets and owned by the community. The compactor and baler could also be used by the community to sell other materials, such as cardboard and other valuable plastic materials, thereby further supplementing incomes.

Scalability and future outlook

Expanding market: Although the communities could not self-finance the shipment of nets after external funding ceased, they have continued to recover, clean, bale and store fishing net waste within their community. TierraMar continues to investigate how the net recycling project can be re-established. For example, since the program ended, the market for clean fishing net waste has grown significantly and buyers now exist within Indonesia and the Asia-Pacific region.

TierraMar is exploring the financial viability of these local buyers, particularly the associated reduced shipment expenses, to determine whether a self-sustaining system can be implemented for the Merauke communities.

Reducing ghost gear: Since the SeaNet Indonesia program, TierraMar has continued to find solutions to reduce fishing net waste. TierraMar has run best-practice management of fishing gear workshops in Vanuatu and Solomon Islands as part of the Commonwealth Litter Program (CLiP) by the Centre for Environment Fisheries and Aquaculture Science (CEFAS) and is continuing to implement GhostNets Australia, focused on reducing the amount of ghost gear washing up in Australia.

5

A step-by-step guide to knowing your waste

James Baker, The Asian Development Bank, Philippines

Before you can reduce any form of waste, you first need to know about the waste you are dealing with. This is easy to do in your own home, but how do you understand the waste of a whole community? This is where James can help us out. Here, he provides a step-by-step guide on how to identify, measure and understand the types of waste you have in your community. He then provides advice on what to do with this knowledge when it comes to waste management and communicating your findings to governments, investors and technology suppliers.

Effective waste management is becoming a focal activity as countries reduce their environmental impact and increase human health. 'Circular economy', 'recycling', 'waste to energy', 'refuse-derived fuel' (RDF) and 'anaerobic digestion' have become common phrases in international, regional and local discussions. All these initiatives begin with a basic understanding of the waste created in an area.

Communities generate many different types of waste, each with its own specific character and solutions; however, the techniques to measure, understand, communicate and ultimately manage waste are universal.

This guide focuses on municipal solid waste (MSW) generated by domestic households, but the steps described here can equally be applied to industrial, agricultural and liquid wastes. Whether a local government or community is designing and funding its own waste treatment system, asking for technical support from suppliers or seeking funding from external sources, the same basic information is needed.

Step 1. Measuring waste: How much waste do I have to manage?

The weight of waste[A] generated can come from one of two methods:

Method 1: from actual collection records, provided in volume (m³) or weight (tonnes[B]). These figures can be retrieved from a contractor's collection/haulage records or from landfill records, either through vehicle counts or weighbridge records.

Method 2: by calculation. Annual waste generation can be calculated by multiplying the number of people in a municipal area by the average daily waste generation figures[C] for that location by 365:

Annual waste generation = population × waste generation per per day × 365

When first studying waste, it is very useful to collect information using both methods and compare the outcomes. The difference between the actual collection records and calculated waste generation will give a clear guide of potentially how much waste is being generated in the community but not reaching official disposal points. This 'missing' waste may be being collected by informal recyclers, burned by households or dumped into the environment.

A further set of investigations to understand the fate of this 'missing' waste is a valuable step towards understanding a community's waste management challenges and the baseline impact of the community on the environment, as well as for targeting future waste management activities.[D]

Knowing the weight of waste needing to be managed allows the community to understand the scale of the challenge, but also has a direct effect on the best solutions available to manage it. Some industrial systems work well on large tonnages of waste (>500 tonnes per day), whereas other industrial systems are optimised for smaller tonnages (down to 20–30 tonnes per day). For communities with smaller tonnages,

---

A Waste can be measured by weight or by volume. This chapter uses weight, but the same results can be achieved by using volume. Most equipment suppliers and investors will use metric weight figures (kilograms or tonnes).

B Make sure you know which unit of measurement is being used (i.e. imperial tons or (metric) tonnes). The difference is small, but over the course of a year the difference can add up to result in under- or over-capacity in the system.

C Average daily waste generation figures provided in kilograms per person per day vary between countries and between rural and urban areas. The World Bank provides a useful source for this information in their *What a Waste* report (https://datatopics.worldbank.org/what-a-waste/).

D A common error when calculating and measuring waste is not taking full account of water and moisture derived both from the waste and from rainfall.

various technologies and systems have been proven to be successful. Recent digital innovation and the wider access to smart phones in target communities has allowed for the development of app-based solutions to community waste challenges, with Octopus and Recity being excellent examples.

Step 2. Identifying waste: What type of waste do I have?

Once a community has calculated the total weight of waste being generated, the next step is to understand what materials are present in the waste. This process is known as a 'waste characterisation' study or sometimes a 'waste mapping and characterisation' study. This process can be achieved by the community with simple weighing scales and sorting tables, or the service can be provided by a specialist contractor.

The outcomes of the characterisation process are a clear picture of the weight of each type of material in the waste stream. Some areas of the world have waste with a very high plastic content, whereas other areas, particularly rural areas, have waste with a very high organic content. Waste characterisations change during the year, particularly in countries with wet and dry seasons or during periods of national or religious celebrations.

The basic characterisation presented in Table 5.1 is sufficient for gaining an initial understanding of community waste and communicating that information to investors or technology suppliers.

Table 5.1. Template for recording waste characterisation with disposal guidance for different waste groups

Waste type	Percentage by weight	Description	Recycling options	Refuse-derived fuel
Food and kitchen waste		Organic	Suitable for composting or anaerobic digestion	Suitable for fuel after drying
Other organic				
Paper and cardboard			Suitable for recycling	Suitable for fuel
Leather and rubber				
Textiles/clothing				
Plastics		Non-organic		
Metal				Not suitable for fuel
Glass				
Ceramic and stone				
Special			Not suitable for recycling	
Other/residual				

The exact solution for the recycling and disposal of each type of MSW depends very much on the results of this characterisation. The decision process by which a community or government selects the most suitable group of solutions is complex and unique to each situation. However, focusing on effective collection and recycling should be the primary target, with safe disposal through managed landfill or fuel solutions being used only for those materials that cannot be returned to the economy.

By successfully completing Steps 1 and 2, a community will have enough information to establish a waste management baseline and provide the necessary information to communicate effectively with national government, technology suppliers and investors. On their own, these steps only scratch the surface of understanding waste management solutions, setting baselines and addressing the environmental impact of a community's waste management.

The following steps offer some additional activities a community should consider when measuring and characterising their MSW.

Step 3. Understanding waste: What else should I have to know about my waste?

Steps 1 and 2 provide a community with a basic understanding of their MSW, but, as with much in life, MSW is rarely that simple.

The water problem

Even though this chapter discusses municipal solid waste, there is always an issue of water and its impact on MSW. Water affects the weight of waste, contributes to 'missing' waste and directly affects possible waste management solutions.

Water or, more correctly, liquids occur in MSW in two broad forms:

Moisture content: this is the level of moisture held within the material and is most typically used for organic items. For example, a discarded fish or piece of fruit has a moisture content. This increases the weight of waste, but can also decrease during the collection process through squashing or decay.

Free water content (sometimes incorrectly called 'leachate'[E]): this is liquid that has been included in the collection due to rainfall or liquids in waste items (e.g. half-full drink bottles being thrown away). Again, this has an effect on the weight of

---

E 'Leachate' is the correct term for liquids that have leached from the base of a landfill or other long-term waste storage system. The liquid encountered during collection and processing is referred to as 'free water'.

An example of a litter recycling station.

the waste; for example, a waste collection vehicle operating after a rainstorm may collect a significant amount of water, which is included in its weight at the weighbridge. However, when the vehicle empties the waste into the waste management site, the water flows away, resulting in a difference in waste weight.

All waste management activities need to consider both the moisture content and free water content during their design process. If the waste is to be disposed of in a landfill, the amount of liquid will affect the leachate generation of the landfill. If the waste is to be turned into fuel, then it needs to be dried to a certain level.

The recycling conundrum

Every item of material that enters a landfill, is thrown into a river or is burned in the backyard represents lost value to the global ecosystem. All the effort, energy, resources and environmental impact involved in producing that item, whether it is a plastic bag or a piece of fruit, is lost as soon as it is thrown away.[F] If an item is not disposed of safely, then its negative impacts on the environment continue as it pollutes the global ecosystem.

By collecting recyclable materials from municipal waste, some of the costs of production can be retained within the economy. If a plastic bag is reused once, then its environmental cost is reduced by roughly 50%. If plastics are recycled, we can avoid

---

F This concept is known as 'environmental full-cost accounting' or 'life cycle analysis'. These are methodologies by which the environmental costs are added to the economic costs of an item to provide a total figure for its cost to the global environment.

the need for the further consumption of raw materials, oil and gas, and the environmental damage incurred during their extraction, to create a new replacement product.

Metals, plastics and glass are strong candidates for recovery and recycling. Paper and cardboard are widely recycled in some areas of the world, but the high moisture content of MSW in Asia reduces the value of paper to recyclers in this region.

Is RDF the answer?

MSW can offer a valuable source of fuel for the generation of heat and/or electricity. Wet organic fractions can be used in anaerobic digestion systems to generate methane-rich biogas. This can be either burned directly for cooking or introduced into a combined heat and power generator to provide electricity and heat for domestic or industrial use.

The combustible elements of the MSW that are not suitable for recycling can be used in RDF for sale to cement factories or power plants. When considering the fuel potential of MSW, the key factors are moisture content and the energy contained in the waste materials. Plastic has a very high energy content but should be recycled rather than turned into RDF. Wet organic material has a very low energy content, especially if it has not been dried.

Final thoughts

Waste management as a sector and field of study is populated with jargon, acronyms and a multitude of approaches that can discourage communities from addressing their waste management challenges. This chapter has attempted to provide a set of simple steps that will allow a community or local government to measure and begin to understand their MSW in such a way that their findings can easily be communicated to nations governments, investors and technology suppliers.

Although not exhaustive, these steps provide the foundation baseline from which discussions can grow. By repeating these steps periodically, a community can begin to not only understand their starting point, but also measure their improvement over time as we all work together to protect our global environment.

6

Finding seed funding for your project: the blue finance world

Azra Yaqub Vawda, independent consultant, Pakistan

After reading about the incredible programs in this book you are probably feeling pretty inspired. You may even have an idea about how you can transform waste into a commodity in your community. But what's next? How do you acquire seed funding to get your idea up and running?

We asked Azra Yaqub to explain the financial opportunities in our current world and what you need to bring to the table when seeking funding from potential investors.

Entering the finance world and getting to know what opportunities are out there can be confronting. Azra steps you through the concept of blue financing, the current growth in the blue economy, the opportunities that are out there and how to maximise them.

Financing climate change and, in particular, the 'blue economy' (all economic sectors that have a direct or indirect link to the ocean) is challenging, with governments facing ever-increasing demands from infrastructure, livelihoods, jobs and healthcare. The COVID-19 pandemic made this even worse by negatively affecting economic growth, requiring further government expenditure. Thus, there is a growing financing gap, leading to calls for the private sector and alternative financing to step up and help.

Traditional blue financing
Blue initiatives thus far have been mainly initiated and financed by governments and international organisations, with foundations and philanthropy aiding this effort. Such actions also contribute to the United Nation's Sustainable Development

Goals (SDGs), in particular SDG 14 ('*Conserve and sustainably use the oceans, seas and marine resources for sustainable development*') and to the Nationally Determined Contributions of countries, further giving them a 'public sector' feel.

However, as the volume of pollution in our waterbodies (oceans, rivers, lakes) continues to increase, the need for financing is increasing, and attention is being focused on the contributors and manufacturers (the source) of the pollution. Plastics manufacturers are an example of this, alongside tourism stakeholders, for instance. To create a better future, organisations need to reduce their pollution, thereby reducing the need for financing for clean-up efforts, and to substitute their products for environmentally acceptable alternatives. From a financial perspective, there are positive impacts of an unpolluted ocean environment for connected businesses (e.g. fisheries and tourism), which makes for an attractive private capital proposition for governments and publicly funded endeavours.

Instruments of traditional financing

Grants: Grants are typically provided by a government department, corporation, trust or foundation and, on occasion, from international organisations such as the World Bank. Grants are often used in blue financing to fund initial project evaluation, impacts and project preparation. They are typically non-repayable, although zero-interest grant loans are also becoming popular. Most grants fund specific projects and require some level of monitoring and evaluation.

Budgets: Government budgets are the primary source of funding blue initiatives. Spending on blue initiatives can be either through industry budgets or, as is now often seen, specialised SDG or climate budgets and ministries. In specific cases where pollution is of particular economic concern, one may even see departmental budgets funding such initiatives.

Loans: As awareness about climate change grows, specialised government/state bank, commercial and development bank loans are also beginning to finance blue initiatives. However, the risk and return scenarios still make these activities inadequate and expensive in commercial banks. Further work is needed to ensure the attractiveness of blue projects.

Why should private capital be interested and what is preventing it from flowing?

The United Nations has estimated that the global blue economy has an annual economic value of around US$2.5 trillion, equivalent to the world's seventh largest economy.[1] Thus, it is not surprising that the global blue economy is attractive for private capital, such as fund investors, capital markets, insurers, banks and policy makers, as a new source of opportunity, resources and profits.

However, private capital is competitive and focuses on both risks and returns. The reason why there is inadequate flow of such investments currently is largely the lack of 'bankable/investable' opportunities: projects where the risk is mitigated in a way to make it attractive for investors and that generate a sufficient return to justify that risk.[2] If such projects could be found, it could unlock blue finance's obvious vast economic potential. But how do we do it?

Enter innovative financing

Innovative financing could offer solutions to the bankability conundrum in terms of tools to make projects bankable and instruments that could be attractive to investors. Simply defined, these are methods that raise new funds for development or adapt traditional methods using different instruments. Topical examples include green bonds, sustainability funds and coronavirus bonds.

Typically, innovative financing can be broken into two parts: (1) instruments, such as bonds and funds; and (2) mechanisms, such as guarantees, facilities, cofinancing and first-loss tranches.

As can be imagined, the level of investor awareness, type of private capital, sophistication of financial markets and geographical location can all determine the kinds of innovative finance possible. Let us look at a few different instruments and mechanisms.

Instruments

A bond is a fixed-income instrument and constitutes a loan made by an investor to a borrower. Companies, municipalities, states and sovereign governments raise money from the capital markets through bonds, the proceeds of which are used to finance different projects. The borrower typically has to make interest payments throughout the life of the bond, and has to repay the principal amount at maturity.

Funds are vehicles with money that are typically allocated for investments for a specific purpose: pension plans, growth, sectors (healthcare, children, SDGs), mitigating disasters and corporate social responsibility.

Both these instruments have been widely used in the private and public sectors globally, and are beginning to become popular in the blue finance field.

Blue bonds: Blue bonds are a relatively new asset class, building on the success of the green bond market, which raises money from capital market investors to finance climate change projects. Blue bonds look to invest monies into bankable projects in the blue economy. The projects have to be revenue generating to repay the interest and principal of the bondholders, and would typically need some level of government and multilateral bank support (revenues and guarantees). They tend to work in a similar manner as regular bonds, but their proceeds are earmarked to be used or invested in projects that specifically relate to the oceans and, of course, plastic pollution. Money from blue bonds can be put in to generate jobs, economic growth and healthy oceans by investing in fisheries, marine and coastal tourism, coastal pollution, circular economy, marine renewable energy and green ports and shipping.

Blue funds: Awareness in the fund community has led to a growing class of impact investors in both the bond and fund worlds. Blue funds are focused on attracting these investors to blue initiatives, alongside climate sustainability and climate fund investors. With no specific fixed interest payable, the risk in such vehicles tends to be higher than in blue bonds, with a time horizon for investment returns that is likely to mean delayed payoffs.

The blue economy also impacts many other areas, from health (SDG 3) to gender equality (SDG 5). Moreover, it also affects poverty (SDG 1): gross domestic products will be affected in the longer term because, if the oceans keep getting polluted, the source of income for many people in the informal sector (e.g. waste pickers, fishers) and all other small-scale sectors will be affected. Thus, SDG funds are also an important source of financing for the blue economy.

In order to make blue funds more attractive (bankable) to investors, both blue bonds and blue funds need to have additional mechanisms and support, including capacity building and investor awareness, so these instruments can be used more efficiently.

Mechanisms

Credit enhancement: This refers to mechanisms by which the quality or credit of the funds or bonds can be enhanced or improved in a way that reduces risk and attracts investors. Investors are concerned that they may not be repaid, both principal and interest, and that these repayments may not occur in a timely manner. Although increased returns on investments can address part of the concern, they do not address all of it. Further, some investors may be limited in the kind of bond or fund they can invest in. Pension funds, for example, can only invest in higher-'rated' investments for much of their portfolio to safeguard their capital and ensure prudent management of retirees promised retirement benefits. Government ratings, as well as the income generated by the blue investments, may not be adequate to attract investors.

Credit enhancements, typically from development banks and governments, effectively try to bridge that gap to attractiveness. Multilateral banks have been supporting climate change and green bonds for quite some time now and are very active in promoting them. Some of the ways of bridging the gap are through the provision of credit enhancements to make bonds and funds bankable. Addressing the risk of inadequate or untimely income from the blue bond/fund investments, governments can provide income guarantees, viability gap funds and reserve funds to give investors comfort. Development banks, often AAA rated, can offer first-loss guarantees and cofinancing to lend credibility.

Pool funding: Projects in the blue area tend to be much smaller, making them unattractive to larger investors due to the time and effort needed to manage and monitor such investments, as well as the lower returns. Private investors want to look at bigger and more diversified portfolios. Pooled funds, with their roots in the municipal markets of Europe, allow for the 'pooling' of many investments under one vehicle, giving investors the size and diversity needed to make them attractive without the headache of managing them. Such funds were first made very popular in Belgium, Denmark and Norway; they worked well and are now being tested in South-East Asia and Africa.

Challenges of finding investors in this field

General challenges: The field is relatively new to investors, without known frameworks and rules. To overcome the primary challenges to investing in the blue economy, the sector and its stakeholders must:

- define standards and metrics
- develop a good pipeline of projects
- improve the understanding of innovative instrument investments
- align taxes and subsidies and improve economic incentives within the blue economy
- improve data and specialist capacity for blue investments.

COVID-related challenges and opportunities: As noted earlier, blue bonds are a relatively newer asset class. The COVID-19 pandemic has negatively affected both bond and fund investors over its duration, with unprecedented losses from typically 'safe' investments. Thus, investors are currently very cautious and unwilling to invest in new and untested investments.

The pandemic saw an increase in investor caution, but has also created opportunities. There has been a considerable increase in awareness of climate- and health-related issues, and there is a growing increase in investors for this sector, and for the SDGs in general. Both the private and public sectors are involved in raising money for initiatives that tackle global issues, such as pollution. With such sustainable bonds under the spotlight, investments in the blue economy could be advantageous as well.

Conclusion

Although financing remains an important element in the equation to solve blue pollution, bankable projects remain key to this issue. Financing must be used not only to fund, but also to de-risk, support and enhance projects in a way that makes them attractive to investors globally. The timing of this is crucial so as to make use of the current heightened interest of investors in SDGs, climate and pollution due to the COVID-19 pandemic. At the same time, recognising that private sector money is competitive and will flow to those initiatives that are impactful, extra efforts must be made to show the impact of the blue economy on other SDGs.

Additional information

We hope you have found this finance information useful. In addition to the formal finance sector, we encourage you to research the grant opportunities available to you. Many of the organisations featured in this book received seed funding from investing their personal savings or from government grants, industry grant partnerships or a mix thereof.

We encourage you to seek out multiple sources to fund your fantastic idea! The chances are that if one person thinks your idea is a winner, then someone else in the financing and entrepreneurial world will probably think so too. We hope this chapter helps you find them!

References

1   United Nations (2022) *United Nations' Sustainable Blue Finance Initiative: Mobilising Capital for a Sustainable Ocean.* United Nations, Nairobi, <https://www.unepfi.org/blue-finance/>.

2   Asian Development Bank (2017) 'Catalyzing green finance, a concept for leveraging blended finance for green development'. Asian Development Bank, Manila.

7

Designing a scalable program

Trish Hyde and Murray Hyde,
founders of The Plastics Circle, Australia

Trish and Murray built a technology recovery program, PlastX, which you may recall reading about in this book. Using the principle of high replicability with low customisation, Trish and Murray developed PlastX as a scalable solution that harvests plastic value before it enters the waste stream and, in so doing, fills corporates' need for recycled plastic to meet their commitments. Here, Trish and Murray share their journey of scaling up an organisation: some good points and some not so good points.

Although stopping marine debris and scaling a fast-food chain may seem incongruous at first, most people will agree that the fast-food giant McDonald's has cracked the secret sauce in terms of successful scalability. McDonald's has around 40 000 outlets in over 100 countries.[1]

Why them? What made them scalable? It was simple: McDonald's devised, tested and refined their early operations to create a process template that was replicated across America and then beyond. However, the process template that dictated operations, production and services did not dictate the menu. Rather, the McDonald's menu is highly customisable to accommodate differences in food supply and cultural preferences according to country; for example, the American apple pie is a taro pie in China.

What has that got to do with developing a scalable program that stops marine litter? Everything! You can build the best pollution diversion program, but if you can't cost-effectively replicate it, while accommodating essential market differences, you are unlikely to get funding to scale.

Design thinking for success and scalability

We are massive fans of design thinking – the iterative, non-linear process of innovation that sees prototyping happening along the development phase, with the focus on the whole journey, not just on the finished product.[2]

From the outset we wanted to find a commercially viable, innovative way to limit the amount of plastic getting into waterways in Asia. This was a huge challenge that could be attacked in many ways! We chose the circular economy route.

Our 'user' is an ecosystem of users, with everyone from the value chain. Based on our corporate careers, we thought we knew a lot; however, outside the corporate structure, we quickly realised there were pockets of users we needed to understand better.

We escalated our immersion process by curating a global series of plastic circular economy, whole-of-value-chain, business forums in Australia, Malaysia, Fiji, Netherlands and Thailand (the latter with the United Nations Environment Programme). We built a network of over 3000 people, from corporations to recyclers to entrepreneurs, and, because of the style of the forums we held, we had candid conversations with everyone.

Making sense of all the new-found information was harder. Our first and most important breakthrough was that circular economy, especially in Asia, was an aspiration yet to be realised. It had to be. Otherwise, why would Asia be awash with plastic waste while brands were still unable to secure adequate supply of quality post-consumer recycled plastic? This focused our innovation space on how to capture the value of used plastic to fulfil unmet recycled plastic demands.

Ideation

Understanding the users, we approached the question from each of their perspectives. Brands wanted clean, quality supply at best price, and this was what their suppliers wanted to provide. Those collecting plastic were generally informal workers, making a subsistence living, with no control over the whole supply chain process. Between the collectors and end users were layers and layers of middlemen, which made the supply chain inefficient and focused on easy resale as opposed to optimal value recovery.

As the ideas came, we applied a very simple filter. We asked ourselves whose interest would our idea serve. Could it work? Could it work commercially? Answers to the last two questions were obvious, but the first question should not be overlooked. For every idea, we asked who we were affecting. How much and how would they react? We then asked who should pay. Were they able to pay? And, importantly, did they have a need for change?

Testing

At the start we mentioned our first pilot, but this was not our first test. Every opportunity we could, we tested to validate or refine every aspect of what would become our model.

There were false starts: a client who committed but subsequently realised they didn't have internal alignment; another potential customer who, as it transpired, really only wanted greenwashing (wanting to look environmentally friendly without putting in the effort).

What do you do? For us, these (and others) were key validation points, as well as providing new inputs for understanding the user and new ideas to achieve the vision.

Determination and conviction

There is another crucial element for successful program scalability. Call it conviction, tenacity, persistence or contrarianism, it is what propels us to take another step when every muscle says stop.

Throughout our immersion process, we constantly heard that we would not succeed: that plastic waste is worthless; that the recycling industry did not have capacity; that we could not compete on price with the existing supply chain; and that informal workers could not use technology. There were even objections based on our race.

We needed to filter the messages we received and decipher their intent. Sometimes it was ignorance, sometimes self-interest and sometimes pure prejudice. This was where conviction came in. Our pilot project in India did not take place because everyone supported it, it took place because we needed proof points to enable us to overcome the naysayers and move forward. We proved that informal workers could embrace technology and be incentivised to collect plastic to value-adding specification. We proved that we could deliver physical and digital chain of custody. Importantly, we proved that our model of commercial partnerships, the template process for our scalability, worked. Having put those objections to bed, we now needed to move on to test unit economics.

Funding: the barrier to scalability

Having a replicable business model template that accommodated location and cultural differences was just the beginning. More funding was required, and early-stage impact funding (equity and grants) was in short supply, especially in Asia.

Some entrepreneurs are fortunate. If you live in the USA or Western Europe, impact investing is more mature and supported by environmental, social and governance investment principles. These regions have a higher risk appetite when it comes to early-stage businesses. Other entrepreneurs may be fortunate enough to have 10 family and friends willing and able to contribute funding. However, for the rest of us, funding can be elusive. Here's the catch: you'll need traction before you can get funded, and you need funding to get traction!

What we have learned along the way (thanks to some wonderful, insightful mentors) is to take small bites and grow your support base. There is a big difference between how much money you need to fully operationalise and how much money you need to assess the next milestone. We did not learn this quickly. And although we would love to have a big pool of money to just get on with it (and divert plastic *en masse*), we know that small bites are more digestible for investors and grant makers. How did we learn this? Through our amazing supporter base, people who saw our vision and who care. We have been fortunate to have many supporters from all over the world.

How to recognise success ... or the other thing

Alongside all the other important stuff, it is important to know what success looks like, and that rarely involves fist pumps and champagne. In our experience, the small milestones are the way-finders that signal success or a time to recalibrate.

A good example of this was our search for that all-important traction point, our first customer. Because our potential clients were big multinational corporations, there was a long lead time. We had one company express interest early in our journey, and we worked for a year trying to close a deal with them. Unfortunately, we were too early: that company had not yet recognised the commercial and regulatory pressures regarding recycled plastic content that would come to bear. We failed. We recalibrated. We refined. And we moved on. Ultimately, we found the right fit. We succeeded. We used the success as motivation and set the next goal ('piece to bite off').

Entrepreneurialism is hard and there are days when you want to scream, times when you feel like you are not succeeding. And then, out of the blue, you have a random conversation and see someone seriously in awe because they see the successes you have forgotten. They see that you are tackling one of the largest global problems, that you have developed an innovative technology/process solution that could be commercially replicable across Asia and that you have taken a risk and successfully built a pivotal pilot in India (remotely and during the pandemic) from the ground using minimal finances. They see that you are doing something to make a better planet.

A balloon washed up on a beach in Australia.
PHOTOGRAPH: KATHRYN WILLIS.

Key messages

Use design thinking: Understand the users within the ecosystem you want to change. Make sense of the information you find and align concepts and people to inform your ideation. Generate ideas and filter them for user resonance, workability and financial viability. But don't reject ideas lightly, and never because one person doubts you. The ultimate decider/refiner is in the testing.

Design for scalability from the beginning: What is it that you are doing (technology, process, product) that can be templated for easy, efficient duplication while accommodating local needs?

Be courageous and persistent: Have a big vision and take at least one step every day. Embrace the idea that you don't know what you don't know. If you did know everything, you probably wouldn't start.

Stay the course: You will hear the word 'no'. A lot. Funders, potential backers and investors say it most often. It is not personal. There are a lot of other early-stage start-ups looking for funding too and only a small pool of funders. Build with what you have and take small steps forward.

Set micro goals: Setting micro goals helps you see and communicate your progress, and breaks down your funding needs into smaller, digestible buckets.

Share: Connect with like-minded people committed to solving plastic pollution challenges. These people can be your greatest supporters and ambassadors. You may also be able to help them achieve their own goals. Let's face it, the plastic pollution problem is big enough for everyone to want to solve.

References

1   McDonald's (2022) *McDonald's reports fourth quarter and full year 2021 results.* [Press release] McDonald's, Chicago, <https://corporate.mcdonalds.com/corpmcd/en-us/our-stories/article/press-releases.Q4-2021-results.html>.

2   Liedtka J (2018) Why design thinking works. *Harvard Business Review* **September–October**, <https://hbr.org/2018/09/why-design-thinking-works>.

8

Closing thoughts

Chris Wilcox, Minderoo Foundation, Australia, and
Kathryn Willis, CSIRO, Australia

The inspiration for this book grew from two experiences in the marine debris team at CSIRO. The first happened walking along a street in central Jakarta, the capital of Indonesia. Looking off a bridge, one of us noticed a man working on a concreted area, next to the river. The riverbank had been converted into a tall concrete wall and the man had built a slender bamboo platform reaching out over the river. Below the platform, sitting on the water surface was another bundle of bamboo, attached to the shore and pointing slightly upstream under the boom. The man had a net on a long pole and was using the lower bundle as a litter boom to collect material floating in the water. He had a large pile of polyethylene terephthalate (PET) bottles up on the concrete, above the river, that he had collected, presumably to sell to a recycler. At the same time, he had a second pile of all the other plastic that he had pulled out of his litter boom sitting in an enormous pile next to his PET pile. Looking at that scene begs the question, is there a way to help him and, in the process, capture all the plastic that he cannot sell, preventing it from following the river out into the sea? Could providing a bit of logistical support, such as bringing a truck that could collect both piles and pay him for the PET, leave everyone better off (no transportation effort for him, no plastic left in the river for the rest of us and a more reliable delivery of PET for recyclers)?

Local ingenuity along the Ciliwung river in central Jakarta. A waste picker has built his own boom, from which 'waste' is scavenged and turned into value.

The second experience occurred at a G20 meeting in Germany, when one of our team was supporting the development of a G20 resolution on addressing the plastic pollution problem. Over the course of 3 days, clauses for the resolution were debated. There was much to-ing and fro-ing among countries about committing resources, the need to improve waste collection systems, the critical need for moving from uncontrolled waste and dump sites to landfills and, in particular, commitments from wealthy countries to provide financial support for these efforts. At the time Indonesia was in discussion with donors to fund waste collection and disposal in 20 cities, with an estimated cost of US$4.5 billion. As the distinguished delegates debated, the complete lack of awareness of the role of the informal sector became increasingly clear. Although all the delegates were committed to resolving the problem, there was no awareness of the role of the informal sector in the collection of waste in most of the world. Doing plastic pollution surveys around the globe over the past decade, we have seen the informal sector in action everywhere, from the poorest areas outside Lima, Peru, to the wealthiest suburbs of Los Angeles, in the USA. All around us, every day, people earn money collecting what others discard; and these people respond to the value of the material, almost instantaneously. Instead of taking a decade to build waste infrastructure, which will likely cost as much to run as to build, we could instead inject a much smaller amount of money into supporting the informal sector, helping these people to collect material more easily and transport it more effectively. After much discussion, the delegates incorporated a statement on the informal sector into the G20 resolution, amidst a host of promises to invest large sums of money into the formal waste collection and recycling sectors.

These two experiences, the firsthand observation of the potential of the informal sector to contribute to solving the plastic pollution problem and the direct experience of the lack of awareness of the role and promise of the informal sector in turning plastic pollution into a valuable commodity, were the genesis for this book. Our goal has been to provide simple summaries, listing all the necessary components and resources, and walking through the steps involved in taking a solution from an idea to an operational system.

Our keen desire in pulling the book together was to highlight the incredible diversity and array of grassroots solutions that we were learning about through working on plastic pollution. We realise that the approaches we have shared here are not comprehensive; there is other important work occurring in other parts of the world, with a suite of entrepreneurs identifying innovative ways to create solutions within their contexts. We plan to continue to collect additional stories of new projects springing up around the world. That said, we hope that ultimately we develop and use better closed-loop systems to design, make, use and circularise plastics products so that we are treating plastic as a commodity rather than waste.

In this book we have focused on stories that looked at bottom-up solutions. This means we really targeted those enterprises that took pieces of a system to grow, produce and develop something more complex from something that was simpler. These are entrepreneurs who have taken creative approaches to make something new, novel and scalable from products that were otherwise treated as waste. Some of these programs created new markets, some linked informal waste pickers to buyers to increase transparency and improve livelihoods. Some programs included here focused strongly on generating revenue, whereas others may have had a stronger focus on repurposing plastic 'waste'.

Some cases we present tap into existing systems as a key to their success: GreenHub in Vietnam taps into an organised social structure of women's groups within the country and, in Brazil, BVRio works within an extended producer responsibility structure. Some of the programs included here are less focused on generating revenue than they may be on repurposing plastic 'waste'. Others may have a limited market, although we aimed to include those programs that have a clearly identified scalable market in which to grow.

All in all, we hope the stories we present inspire other innovators, activists and everyday people to think about strategies to address the plastic pollution problem we face. And where we can bring these bottom-up solutions that build community, give people a sense of purpose and generate new livelihoods for participants to decision makers, we can move from the world we have to the one in which we would like to live. And if you are a G20 environment or infrastructure minister, you can smile at where a bit of indignation takes one – please, open your mind to the beautiful world of bottom up solutions.

Thanks for reading!

INDEX